Peter Buhler

Nanothermodynamics

Peter Buhler. Nanothermodynamics: Translated by Dr. V.Ushakov – St.Petersburg, "Yanus", 2004. 170 p.
ISBN – 5-9276-0042-5, ISBN – 3-7896-0676-6

The book is a continuation of the author's following books: Physicochemical Thermodynamics of Substance (2000), Thermodynamics of substances at high pressures (2001), published by Yanus Publishing house, St. Petersburg.

The thermodynamic properties of nanodispersed substances, nanodispersed solutions, and colloidal dispersed systems have been discussed in the book. The following well-known phenomena have been quantitatively described: an increase in the saturated vapor pressure of nanodispersed substance, a decrease in the saturated vapor pressure of liquid nanophase at the surface of solid substance (physical adsorption), melting temperature depression for nanodispersed substance, and other features of substance in the nanodispersed state.
The book can be ordered (30 EURO for a book) at e-mail (yanus_ph@rambler.ru).
This book can be ordered at the publisher Verlag Palm & Enke, Erlangen, ISBN 3-7896-0676-6.

The author is grateful to Helene A. Buhler for the prepress of the manuscript in Russian.

Книга отпечатана по материалам, предоставленным автором.

ISBN - 5-9276-0042-5
ISBN – 3-7896-0676-6

CONTENTS

Introduction

At present, the branches of the general science dealing with the condensed substance of nanosize (1÷100 nanometer) have been given the name of *nanoscience.* Thus, *nanophysics and nanochemistry* study physical and chemical properties of nanodispersed substance. Therefore the author has named the division of thermodynamics studying thermodynamic properties of nanodispersed substance *nanothermodynamics*. As against the general thermodynamics, in which the dependence of thermodynamic properties of substance (for example, internal and free energy) on temperature and pressure is studied, *nanothermodynamics* considers the influence of the surface area of substance on its thermodynamic properties in addition to that. Therefore the main task of *nanothermodynamics* is to study the influence of the surface energy of substance on its thermodynamic properties and its interaction with other substances or aggregative states of a given substance.

The laws of thermodynamics at isobaric-isothermal increase in the surface area of substance are set out, and the following phenomena are quantitatively described in the book:

– increase in the saturated vapor pressure of a nanodispersed substance,
– depression in the saturated vapor pressure of the nanophase of a liquid at the surface of a solid (physical adsorption),
– capillary flow of a liquid,
– depression in the melting temperature of a nanodispersed substance,
and also
– thermodynamic properties of colloidal dispersed systems,
– equilibrium of chemical reactions involving nanodispersed substances,
– change in the internal energy of a substance at its transition from the solid state to atoms or molecules in the gas phase through the nanodispersed state.

The book is a continuation of the author's following works: Physicochemical Thermodynamics of Substance (2000), "Fiziko-khimicheskaya termodinamika veshchestva" (2001), "Thermodynamics of substances at high pressures" (2001), and "Termodinamika veshchestva pri vysokikh davleniyakh" (2002) published by Yanus Publishing house, St. Petersburg*.

Peter Buhler
January, 2004, Erlangen, Germany
(e-mail:peter.buhler@gmx.de)

* The listed books can be ordered (20 EURO for a book) at e-mail: peter.buhler @gmx.de

SYMBOLS AND DEFINITIONS

Name	Symbol	Definitions	Dimensionality
		Pure Substance	
Cubic expansion coefficient	α	$\alpha = (1/V)(\partial V/\partial T)_P$	K^{-1}
Isothermal compressibility	k_T	$k_T = -(1/V)(\partial V/\partial P)_T$	Pa^{-1}
Heat	q	Energy transferred from one substance to another	J
Thermal coefficient of heat	C_p	$C_p = (\partial q/\partial T)_P$	$J \cdot K^{-1}$
Pressure coefficient of heat	β_q	$\beta_q = (\partial q/\partial P)_T$	$J \cdot Pa^{-1}$
Work	w	energy lost by substance at its expansion or gained by substance at its compression	J
Thermodynamic temperature	T		K
Celsius temperature	t	$t = T - 273.15$	°C
Pressure	P		Pa
standard pressure	P^0	$P^0 = 10^5$ Pa (1 bar)	Pa, bar
relative pressure	f	$f = P/P^0$	1
Internal energy of substance	U		$J \cdot mol^{-1}$
absolute level at $T = 0$ K	U_0	$U_0 = 0$	$J \cdot mol^{-1}$
increment in the internal energy of a substance relative to U_0 or the internal energy of substance	U		$J \cdot mol^{-1}$
Thermal coefficient of the internal energy of substance (energy capacity)	C_U	$C_U = (\partial U/\partial T)_P$	$J \cdot K^{-1} \cdot mol^{-1}$
Pressure coefficient of the internal energy of substance	D	$D = (\partial U/\partial P)_T$	$J \cdot Pa^{-1} \cdot mol^{-1}$

Entropy of substance	S	$dS = dU / T$	J·K^{-1}·mol^{-1}
absolute value at T = 0K	S_0	$S_0 = 0$	J·K^{-1}·mol^{-1}
absolute value at a temperature of T	S_T	$S_T = \int_0^T \frac{C_U}{T} dT$	J·K^{-1}·mol^{-1}
Free energy of substance	G	internal energy level of substance is TS less than U: $G = U - TS$	J·mol^{-1}
absolute level at T = 0K	G_0	$G_0 = U_0 = 0$	J·mol^{-1}
decrease in the free energy of substance relative to G_0 or the free energy of substance	G	$G = G_T - G_0$	J·mol^{-1}
Thermal coefficient of the free energy of substance (entropy of substance)	S	$S = -(\partial G / \partial T)_P$	J·K^{-1}·mol^{-1}
Pressure coefficient of the free energy of substance	E	$E = -(\partial G / \partial P)_T$	J·Pa^{-1}·mol^{-1}

Nanodispersed substance

Molar surface area	A		m^2·mol^{-1}
Internal energy of nanodispersed substance	U^d	$U^d = U + U^S$	J·mol^{-1}
Internal energy of bulk substance	U		J·mol^{-1}
Surface energy	U^S	$U^S = \sigma A$	J·mol^{-1}
Specific surface energy	σ	$\sigma = \left(\frac{\partial U^S}{\partial A}\right)_{P,T}$ $\sigma = -\left(\frac{\partial G^S}{\partial A}\right)_{P,T}$	J·m^2
Free energy of nanodispersed substance	G^d	$G^d = G + G^S$	J·mol^{-1}
Free energy of bulk substance	G		J·mol^{-1}

Surface free energy	G^S	$G^S = -\sigma A$	$J \cdot mol^{-1}$
Entropy of nanodispersed substance	S^d	$S^d = -\left(\frac{\partial G^d}{\partial T}\right)_{P,A}$ $S^d = S + S^S$	$J \cdot K^{-1} \cdot mol^{-1}$
Entropy of bulk substance	S	$S = -\left(\frac{\partial G}{\partial T}\right)_{P,A}$	$J \cdot K^{-1} \cdot mol^{-1}$
Surface entropy	S^S	$S^S = -\left(\frac{\partial G^S}{\partial T}\right)_{P,A}$ $S^S = s^S A$	$J \cdot K^{-1} \cdot mol^{-1}$
Specific surface entropy	s^S	$s^S = -\left(\frac{\partial \sigma}{\partial T}\right)_{P,A}$	$J \cdot K^{-1} \cdot m^{-2}$
Pressure coefficient of the internal energy of nanodispersed substance	D^d	$D^d = \left(\frac{\partial U^d}{\partial P}\right)_{T,A}$	$J \cdot Pa^{-1} \cdot mol^{-1}$
Pressure coefficient of the free energy of nanodispersed substance	E^d	$E^d = -\left(\frac{\partial G^d}{\partial P}\right)_{T,A}$	$J \cdot Pa^{-1} \cdot mol^{-1}$

1

ON THE BASIC CONCEPTS OF CLASSICAL THERMODYNAMICS

At the middle of the nineteenth century, namely, when the law of conservation of energy was worked out in 1842, the intensive advancement of thermodynamics began, and the concepts as internal energy, heat, and work were defined. At the same time, no account was taken of the fact that the internal energy is a property of substance, while heat is not. The basic concepts of classical thermodynamics considered in the chapter are not realized in Nature for it exists under the laws which are irrespective of anybody. The author's books [8, 9] have been used when presenting the first chapter.

1.1. Internal energy, heat, and work

1.1.1. Internal energy is a property of substance, while heat and work are types of energy transfer

It follows from experience that at given temperature and pressure a substance possesses the certain internal energy U, and the change in it is a measurable quantity (with respect to the value $U_0 = 0$ at $T \rightarrow 0\text{K}$). Therefore the internal energy is a property of substance (or a state function). An increase in the internal energy of a substance occurs due to heat input q or the work of compression. In both cases an increase in internal energy dU follows the first law of thermodynamics.

If at constant pressure a certain amount of heat dq is delivered to a substance the internal energy of the substance increases by dU, and the substance does the work of expansion dw against the surroundings at the expense of dq. The following energy balance between dq, dU and dw:

$$dq = dU + dw \tag{1.1}$$

is the first law of thermodynamics at constant pressure. As the internal energy is a property of substance equation (1.1) may be written as follows:

$$dU = dq - dw\,. \tag{1.2}$$

If a substance is subjected to compression at constant temperature, i.e. the certain work dw is done, the internal energy of the substance will increase by dU^P, and the certain amount of heat dq will be released into the environment. The following energy balance between dw, dU^P, and dq :

$$dw = dU^P + dq \tag{1.3}$$

is the first law of thermodynamics at constant temperature. Taking into account, that dU^P is a characteristic of substance, equation (1.3) may be written as follows:

$$dU^P = dw - dq. \tag{1.4}$$

In equations (1.1) and (1.2) the heat dq is the total energy supplied to the substance from the environment, and the work dw is a part of energy dq which is removed to the environment. In equations (1.3) and (1.4) the work of compression dw is the total energy delivered to the substance from its environment, and the heat dq is a part of energy dw which is released into the environment.

From the above it appears that a substance does not possess heat q or work w, as differentiated from the internal energy U. Consequently, these quantities are not properties of a substance (functions of state).

Heat and work are two types of energy transfer.

This is to say that the temperature derivative of heat is not a property of a substance. Nonetheless, heat is used in classical thermodynamics for defining so-called heat capacities of substance: at constant pressure (C_P) and at constant volume (C_V):

$$C_P = \left(\frac{\partial q}{\partial T}\right)_P, \tag{1.5}$$

$$C_V = \left(\frac{\partial q}{\partial T}\right)_V. \tag{1.6}$$

As heat q is not a property of a substance, equation (1.5) does not characterize the substance but only the heat supplied to the substance. The stability of volume required by equation (1.6) is not generally realized by Matter; consequently, the equation can be described as an unrealizable request. This is true for all equations of classical thermodynamics in which the volume stability for the condensed substance is required.

To find the temperature coefficient of internal energy of substance C_U the first law of thermodynamics at constant pressure should be used, i.e. equation (1.2) from which follows:

$$\left(\frac{\partial U}{\partial T}\right)_P = \left(\frac{\partial q}{\partial T}\right)_P - \left(\frac{\partial w}{\partial T}\right)_P. \tag{1.7}$$

As $dU = C_U dT$ and $dw = PdV = \alpha PVdT$ at P = const, we find the following in view of equation (1.5):

$$C_U = C_P - \alpha PV. \tag{1.8}$$

In this equation C_U is the energy capacity of a substance, C_P is the heat capacity, α and V are the cubic expansion coefficient and the molar volume of a substance respectively.

At sufficiently high pressure the value of αPV makes up more than 10% of C_P (see section 2.1). At high pressure it is therefore necessary to use C_U instead of C_P [8, 9].

1.1.2. Transferring energy from a substance to another one is irreversible

As is known, transferring energy from a substance to another one is irreversible. That is the essence of the first law of thermodynamics from which it follows (see equation (1.1)) that at P = const part of the energy transmitted to substance as the heat dq passes to the environment as the work of expansion dw.

Despite the fact that the irreversibility of transfer of energy is built into matter and is well-known, classical thermodynamics uses the concepts of "reversible heat" and "reversible work". The limitation about the fact that the concepts refer to indefinitely slow processes only, is not an argument as the laws of thermodynamics do not depend on time. The concept of "reversible heat" is discussed in paragraph 1.2.2.

1.1.3. The internal energy of substance does not increase owing to the product PV and in consequence enthalpy ($H = U + PV$) is not a property of substance

The idea of the generating energy according to the particular notions of a person has cherished his imagination during centuries. The flow of the continuously operating machine projects by ambitious scientists was so great that in 1775 the French Academy of Sciences took a decision not to examine the designs of "Perpetuum mobile". That this was a right decision was confirmed by the discovery of the law of conservation of energy in 1842, according to which energy does not originate from anything and does not disappear, but is transformed at physical and chemical interaction of substances. Thus no design of perpetual motion can be realized because the law of conservation of energy is fulfilled in the Nature. But the idea of "Perpetual motion" was introduced into science contrary to this universally known law.

In 1876 American mathematician J.W. Gibbs coined the term of enthalpy (H) defined by the following equation:

$$H = U + PV, \tag{1.9}$$

where U and V are the internal energy and the volume of a substance respectively, and P is the ambient pressure.

The equation is quite correct mathematically, as the product PV has the dimensionality of energy. However the internal energy of a substance cannot be raised thanks to our skill to multiply the volume V by the ambient pressure P and to get a certain number with dimensionality of energy. The mathematical correctness of equation (1.9) is a corollary of mathematics as a science, within which "we do not have to say what we are talking about" (R. Feynman).

As a matter of fact, equation (1.9) mathematically expresses the possibility to generate energy from nothing. With regard to the aforesaid, enthalpy is not a property of substance and the concept "enthalpy" is not used in this book.

1.1.4. Internal energy of substance increases with the growth of pressure

As mentioned above, the volume of a substance decreases at the isothermal growth of pressure, and the work of compression dw is expended for both an increase in the internal energy dU^P of a substance and the heat dq transferred to the surroundings:

$$dw = dU^P + dq. \tag{1.10}$$

As the heat transferred to the environment does not exceed the work of compression that has been done, from equation (1.10) follows $dU^P > 0$, i.e. the internal energy of a substance increases with the growth of pressure.

Contrary to this well-known fact the contrary assertion is established in classical thermodynamics, i.e. that the internal energy of a substance decreases with the growth of pressure (see section 2.1). The conclusion follows from equation (1.9) describing the generation of energy due to the product PV.

Fig. 1.1 (curve 1) gives the pressure dependence of the internal energy of water at 100 °C, calculated in Ref.30 by the equations of classical thermodynamics. For reference, the same pressure dependence of the internal energy of water at 100 °C derived here with the help of the first law of thermodynamics and the experimental data from the reference-book mentioned above [30] is also shown in Fig.1.1 (curve 2).

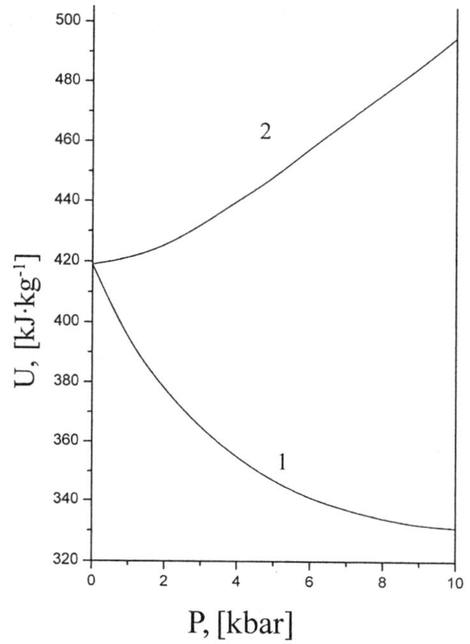

Fig. 1.1. Pressure dependence of internal energy for water at 100 °C: 1 – on the data [30], 2 – calculation on the experimental data [30] using the first law of thermodynamics.

A comparison between curves 1 and 2 in Fig.1.1 shows that the internal energy of water with the growth of pressure increases according to the first law of thermodynamics (curve 2) and does not decrease (curve 1). A decrease in the internal energy of water at isothermal compression would mean that the amount of the heat released into the environment exceeds the work of compression that has been done. It is well-known that matter does not realize the interchange of energy set by classical thermodynamics.

It is interesting to note, that the calculations of the *U,P* curves for water are made using the equations set developed by the International association for the properties of water and water vapor. The last version of the equations set for calculation of thermodynamic properties of water and water vapor was developed in 1997 at IAWS [Industrial Formulation 1997 for the Thermodynamic Properties of Water and Steam. International Association for the Properties of Water and Steam]. It is astonishing, that for already many dozens of years the above international association has been developing equations set which neither water nor any other substance follows. The thing is that the behavior of substance is determined by the first and second laws of thermodynamics, instead of the defining equation (1.9) for enthalpy and other equations deduced using it.

1.2. Free energy and entropy of substance

1.2.1. Free energy is the level of internal energy and therefore free enthalpy ($G^G=H-TS^G$) is not a property of substance

In addition to internal energy U, substance possesses the level of internal energy G located below U by the product TS:

$$G = U - TS. \qquad (1.11)$$

In this equation, S is the entropy of substance which is defined by the following equation:

$$S = -\left(\frac{\partial G}{\partial T}\right)_P. \qquad (1.12)$$

With (1.12) taken into account, from (1.11) follows:

$$dU = TdS. \qquad (1.13)$$

Equations (1.11) and (1.13) are the mathematical expression of the second law of thermodynamics in the integrated and differential forms.

A change in two levels (U and G) of internal energy for nickel with increasing temperature from 0 up to 300K is shown in Fig. 1.2. The curves U,T and G,T come from the same point ($U_0 = G_0 = 0$) and are spaced TS apart according to the second law of thermodynamics.

As previously mentioned, in classical thermodynamics there is another level of energy $H=U+PV$ along with the internal energy U. H is termed enthalpy.

Free energy G^G corresponding to enthalpy H, is called Gibbs energy or free enthalpy [32] and is defined by the following equation:

$$G^G = H - TS^G, \qquad (1.14)$$

where S^G is defined as follows:

$$S^G = -\left(\frac{\partial G^G}{\partial T}\right)_P. \qquad (1.15)$$

With (1.15) taken into account, the mathematical expression of the second law in classical thermodynamics follows from (1.14):

$$dH = TdS^G. \qquad (1.16)$$

As noted above (section 1.1.3), enthalpy is not a property of a substance. Consequently, free enthalpy G^G and entropy S^G, which were introduced by means of enthalpy, also are not properties of a substance.

However at low pressure the product PV can be neglected in equation (1.9), and then at $P \to 0$ we find: $H \approx U$, $G \approx G^G$, $S^G \approx S$.

In some cases so-called Helmholtz energy F is used in classical thermodynamics. It is defined in the following way:

$$F = G^G - PV. \qquad (1.17)$$

In this equation Gibbs energy G^G and the product PV are not properties of a substance, and consequently, F cannot be a property of a substance, and it is not used in the book. The value of PV in equation (1.17) can be neglected at low pressure ($P \to 0$), then we obtain: $F \approx G^G \approx G$.

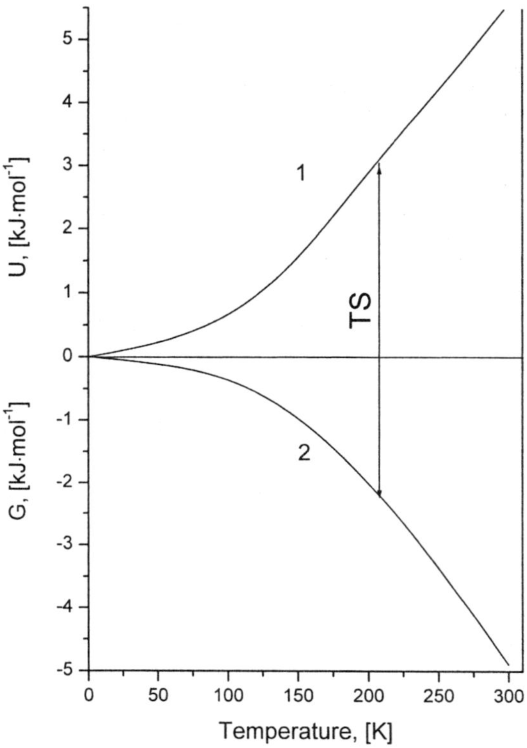

Fig. 1.2. Temperature dependence of internal (curve 1) and free (curve 2) energies for nickel.

1.2.2. Entropy is a property of substance and not that of "reversible heat"

The concept of entropy S has been introduced in 1865 by Rudolf Clausius [1] in the form of the defining equation:

$$dS = \frac{dq_{rev}}{T}, \qquad (1.18)$$

where q_{rev} is reversible heat. Equation (1.18) still is the defining equation of entropy at the present time [32]. The law named the second law of thermodynamics is formulated on the basis of equation (1.18) in Ref. [1]:

Entropy of the universe tends to a maximum.

Recent wordings of this law by way of the statement that “the entropy of the universe increases as a result of physical processes occurring in it until the maximum will be achieved”, does not differ from Clausius’s statement. Authors of all present statements of the second law of thermodynamics are united by the absence of facts in favor of assertion about the continuous increase in entropy of the universe. These facts are not at hand at present yet as well as in Clausius’s time.

Let us now turn to equation (1.18). Any researcher who is studying it is at a loss for two reasons: firstly, reversible heat does not exist and, secondly, heat is not a property of a substance, in contrast to entropy.

Therefore entropy needs to be expressed as the property of a substance using other property of a substance, namely internal energy with the help of equation (1.13) which can be written by analogy with (1.18) as follows:

$$dS = \frac{dU}{T}. \qquad (1.19)$$

The given equation is the mathematical expression of the second law of thermodynamics which becomes in the integrated form as follows

$$U - G = TS. \qquad (1.20)$$

In this equation, entropy S is the thermal coefficient of free energy G:

$$S = -\left(\frac{\partial G}{\partial T}\right)_P. \qquad (1.21)$$

In equations (1.19) – (1.21) U and G are properties of a substance. Consequently, entropy S is a property of a substance, and not that of “reversible heat”. The curve S, T for nickel is shown in Fig. 1.3.

1.2.3. Entropy of substance has nothing to do with thermodynamic probability

According to the second law of thermodynamics (see equation (1.13)) at given temperature entropy of any substance is determined only by the temperature dependence of its internal energy at given pressure. Figure 1.3 presents the temperature dependence of entropy for nickel calculated using the curve U, T (see Fig. 1.2, curve 1). The given curve does not require any treatment, for it has been

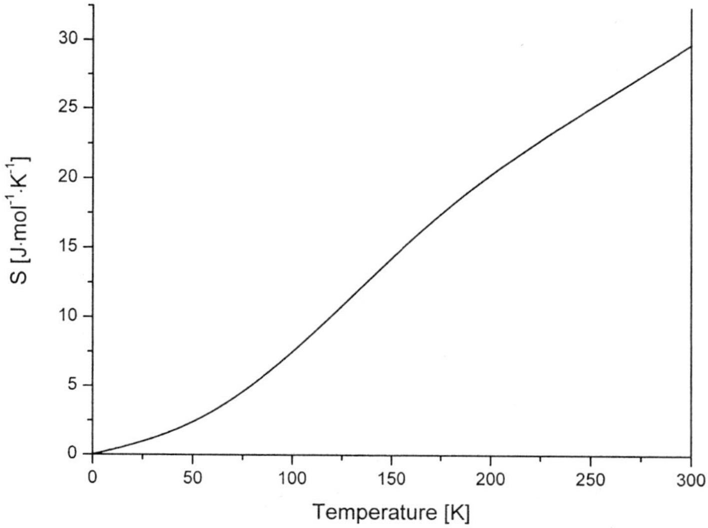

Fig. 1.3. Temperature dependence of entropy for nickel.

constructed using a basic law of thermodynamics, namely, the second law of thermodynamics which is valid for any substance in any aggregative and structural state.

Despite this truism, in 1877 Boltzmann suggested treating entropy as a measure of number of possible microstates *W* which make it possible to realize the given macrostate. In 1901 Maks Plank named the constant of proportionality between entropy *S* and ln*W* by the Boltzmann constant *k*. As a result, he offered the following relation for entropy:

$$S = k \ln W. \tag{1.22}$$

The quantity *W* in the given equation is also referred to as thermodynamic probability.

Thus, classical thermodynamics provides an equation, which completely excludes both experiment and the second law of thermodynamics. Actually a curve of the temperature dependence of entropy of a substance is evaluated at given pressure with the help of the law and an experimental curve *U*, *T*. The curves in a wide temperature interval which covers solid, liquid, and gaseous states for thousands of substances are available in reference-books [20 – 30]. It is important to note that equation (1.22) has never been used for defining entropy of some substance though it is one of the most known equations in science.

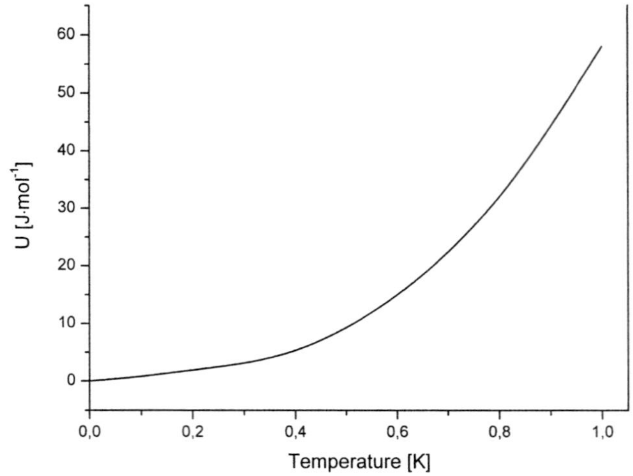

Fig. 1.4. Change in the internal energy of glassy silica in the range of temperatures from 0 up to 1K.

1.2.4. Entropy of any substance equals zero at $T \rightarrow 0K$

According to modern views, only entropy of an ideal crystal is equal to zero at $T \rightarrow 0K$. Entropy of amorphous substances is more than zero at $T \rightarrow 0K$. Modern experimental methods show that all the substances (amorphous and crystalline) lose the capacity to release energy at $T \rightarrow 0K$. That suggests that entropy of all substances is equal to zero at $T \rightarrow 0K$. As an example, Fig.1.4 shows the temperature dependence of internal energy for glassy silica at $T = 0 - 1K$. As may be seen, $dU = 0$ at $T \rightarrow 0K$, therefore according to the second law of thermodynamics (see equation (1.13)) $dS = 0$ and $S = 0$ at $T \rightarrow 0K$, as $T \neq 0K$.

1.3. Pressure on each side of the curved surface is identical

Let us consider a system consisting of a gas phase α and a condensed phase β, with the phase β being a sphere of radius r (Fig. 1.5). If pressure increases from $P \rightarrow 0$ up to some value P at constant temperature, the internal energy of phases α and β will increase by quantities $U^P(\alpha)$ and $U^P(\beta)$, and free energy will decrease by quantities $G^P(\alpha)$ and $G^P(\beta)$.

In this case the second law of thermodynamics in the integrated form becomes as follows (see chapter 2 and [8, 9]):

$$U^P + G^P = EP. \qquad (1.23)$$

In the equation (1.23) $E=-(\partial G^P/\partial P)_T$ is the pressure coefficient of free energy.

In equilibrium the pressure in any point of phases α and β is defined by the following equations:

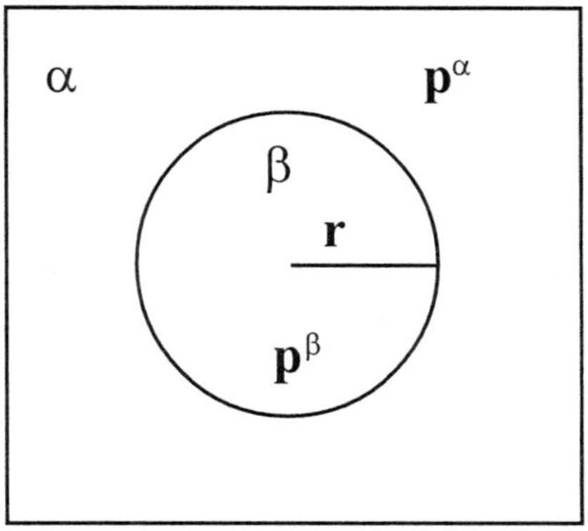

Fig. 1.5. Sphere of radius r from the condensed substance β in equilibrium with the gas α. P^α and P^β are equilibrium pressures outside and inside the condensed substance respectively.

$$P^\alpha = \frac{U^P(\alpha) - G^P(\alpha)}{E(\alpha)}, \qquad (1.24)$$

$$P^\beta = \frac{U^P(\beta) - G^P(\beta)}{E(\beta)}. \qquad (1.25)$$

The second law of thermodynamics is true for any substance and it does not depend on the curvature of phase boundary. Therefore:

$$P = P^\alpha = P^\beta. \qquad (1.26)$$

As this equality is valid for any point of equilibrium system, the following equation

$$P^\alpha - P^\beta = \frac{2\sigma}{r}, \qquad (1.27)$$

which is postulated by classical thermodynamics and where σ is the specific surface energy of the phase β, contradicts the second law of thermodynamics. The difference $P^\alpha - P^\beta$ is named the capillary pressure, and equation (1.27) is accepted as the statement for which there is no experimental data. Therefore equation (1.27) in this work is not used.

1.4. Interface between two condensed phases is established by the chosen phases

In classical thermodynamics a conventional surface is accepted as the interface between two condensed phases α and β instead of a natural interface between them. It is *mentally and arbitrarily passed* inside an, again arbitrarily *chosen* interfacial layer with *unknown* thickness τ. With the interface of phases defined in such a way it completely depends on an arbitrariness of our thought.

Actually the interface between two condensed phases is established by the phases irrespective of any desires and requirements of ours. Therefore an interface which is established by substances and which can be determined is accepted here as the interface between two condensed phases.

1.5. Volume of substance is not the pressure derivative of free energy

In classical thermodynamics, it is customary to believe contrary to experimental data, that the volume of a substance is defined as the pressure derivative of free energy at constant temperature. The defining equation for volume in classical thermodynamics can be written at $P \to 0$ and $G = G^G$ as follows:

$$V = \left(\frac{\partial G}{\partial P}\right)_T . \quad (1.28)$$

As is known, internal and hence free energy of a substance is pressure independent at low pressure, i.e.:

$$\left(\frac{\partial G}{\partial P}\right)_{T,P\to 0} = 0. \quad (1.29)$$

As the volume of a substance at given temperature and $P \to 0$ takes the greatest value instead of zero value, equation (1.28) is fundamentally incorrect. Besides, the internal energy of a substance increases with the growth of pressure, and the free energy decreases according to the second law of thermodynamics [8, 9]. Therefore the defining equation for pressure coefficient of free energy E should be written as follows:

$$E = -\left(\frac{\partial G}{\partial P}\right)_T . \quad (1.30)$$

In the equation the pressure coefficient of free energy E is a measurable quantity which is equal to zero at $P \to 0$.

1.6. Free energy and entropy of mixing an ideal binary solution equal zero

As is known, energy of mixing an ideal solution equals zero $\Delta_{mix}U^{id} = 0$ as in this case at mixing two components either energy does not change. Variation in entropy is due to the change in internal energy, which is equal to zero when an ideal solution is formed. In this case, applying the second law of thermodynamics we have:

$$d\Delta_{mix}U^{id} = Td\Delta_{mix}S^{id} = 0. \qquad (1.31)$$

From this equation follows:

$$\Delta_{mix}S^{id} = 0, \qquad (1.32)$$

i.e. entropy of mixing an ideal solution equals zero, instead of some universal value $(\Delta_{mix}S^{id})_{cl}$ which is calculated in classical thermodynamics by means of the following equation:

$$\left(\Delta_{mix}S^{id}\right)_{cl} = -R(x_A \ln x_A + x_B \ln x_B). \qquad (1.33)$$

According to this equation it is only the concentration of components of a solution x_A and x_B should be preset for one to be able to calculate the entropy of mixing an ideal solution at any temperature. As is known from experience, at any concentration of an ideal solution its energy and entropy of mixing are equal to zero. Since the quantities $\Delta_{mix}U^{id}$ and $\Delta_{mix}S^{id}$ are related to the free energy of mixing $\Delta_{mix}G^{id}$ an ideal solution by the following equation:

$$\Delta_{mix}G^{id} = \Delta_{mix}U^{id} - T\Delta_{mix}S^{id}, \qquad (1.34)$$

with the help of equalities $\Delta_{mix}U^{id} = 0$ and $\Delta_{mix}S^{id} = 0$ we find :

$$\Delta_{mix}G^{id} = 0, \qquad (1.35)$$

not

$$\left(\Delta_{mix}G^{id}\right)_{cl} = RT(x_A \ln x_A + x_B \ln x_B), \qquad (1.36)$$

which is used to calculate the free energy of mixing an ideal solution in classical thermodynamics. Equations (1.33) and (1.36) are not used in the book.

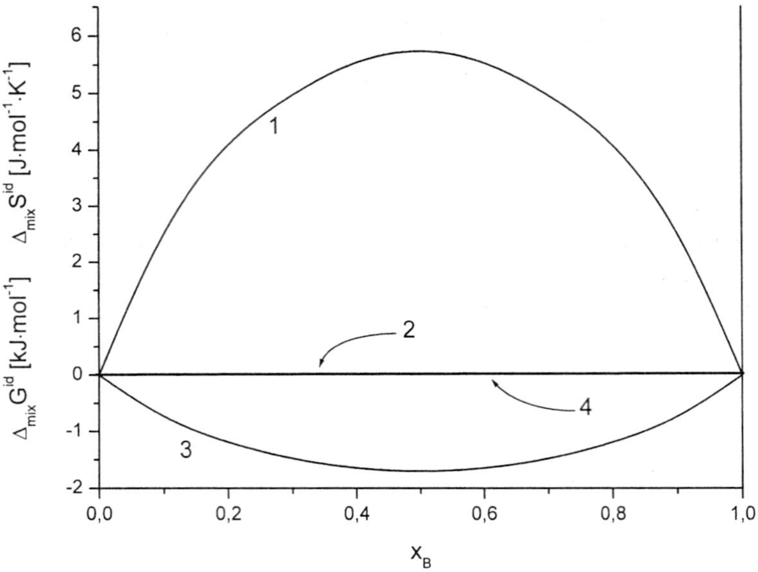

Fig. 1.6. Entropy and free energy of mixing an ideal binary solution as a function of composition: 1– a universal curve $(\Delta_{mix}S^{id})_{cl}$, x_B, calculated by equation (1.33), 2 – curve $\Delta_{mix}S^{id}$, x_B, which corresponds both the experiment ($\Delta_{mix}U^{id}= 0$) and the second law of thermodynamics, 3 – curve $(\Delta_{mix}G^{id})_{cl}$, x_B, calculated by equation (1.36) at 298K, 4 – curve $\Delta_{mix}G^{id}$, x_B, fits both experiment ($\Delta_{mix}U^{id}= 0$) and the second law of thermodynamics.

The curves $(\Delta_{mix}S^{id})_{cl}$, x_B and $(\Delta_{mix}G^{id})_{cl}$, x_B constructed by equations (1.33) and (1.36) (at 298K) are shown in Fig. 1.6 (curves 1 and 3). As evident, the curve 1 has a hump, while the curve 3 has a minimum at $x_B = x_A = 0.5$. Straight lines 2 and 4 correspond to equalities $\Delta_{mix}S^{id} = 0$ and $\Delta_{mix}G^{id} = 0$. They have been derived using experimental data ($\Delta_{mix}U^{id}= 0$) and the second law of thermodynamics.

2

TEMPERATURE AND PRESSURE DEPENDENCE OF THERMODYNAMIC PROPERTIES OF SUBSTANCE

In this chapter, the three laws of thermodynamics have been presented:
firstly, at isobaric increase of temperature;
secondly, at isothermal increase of pressure.
Particular attention has been paid to the physical meaning of heat and work in relation to a substance to which heat is supplied or on which work is done. The following concepts of classical thermodynamics:

– enthalpy,
– reversible heat,
– reversible work,
– entropy as a function of thermodynamic probability,
– volume of a substance as the pressure derivative of free energy,

which do not characterize the real behavior of matter within the framework of the general laws of the Nature, are not used by the author. The questions presented in this chapter are considered in greater detail in the author's works [8, 9].

2.1. The first law of thermodynamics at constant pressure and temperature dependence of internal energy of substance

Let us consider an increase in internal energy of a substance by dU due to heat supplied dq. Experience reveals that $dq > dU$. This is due to the fact that the energy passage to a substance leads to an increase in its volume V, a part of the energy supplied as heat is spent for the work of expansion dw. The energy balance between dU, dq, and dw is the mathematical expression of the first law of thermodynamics at any constant pressure:

$$dU = dq - dw. \tag{2.1}$$

The plus signs of dU and dq in equation (2.1) mean that the internal energy of a substance increases due to heat input. The minus sign of dw means that energy as work releases to the environment. From above it follows that

heat and work are two types of energy transfer.

The equation of the first law of thermodynamics (2.1) is used when deriving the temperature dependence of the internal energy of a substance. In view of relations for dU, dq, and dw which can be expressed as follows:

$$dU = C_U\, dT, \tag{2.2}$$

$$dq = C_P\, dT, \tag{2.3}$$

$$dw = PdV \tag{2.4}$$

or

$$dw = \alpha PVdT \tag{2.5}$$

we obtain the following equations:

$$C_U = C_P - \alpha PV, \tag{2.6}$$

$$dU = (C_P - \alpha PV)dT. \tag{2.7}$$

In these equations C_U – the energy capacity of a substance, C_P – the heat capacity, α – the cubic expansion coefficient, V – the molar volume of a substance, P – pressure.

It is only the values of C_P, describing the amount of heat dq required to raise the internal energy of a substance by dU, that have been experimentally determined. Let us therefore compare the values of αPV for different substances with the values of C_P at various pressures. The results of the comparison for melts of some substances are listed in Table 2.1.

The quantity αPV does not exceed 0.65 % of C_P at a pressure of 100 bar. When the pressure is reduced down to 10 bar the quantities of αPV decrease and

Table 2.1. Values of αPV in equation (2.6) for some melts at various pressures.

Substance (melt)	T [^{0}C]	C_P [J·mol^{-1}·K^{-1}]	αPV [J·mol^{-1}·K^{-1}] at P [kbar] 0,1	1	10	P>3*
Ga	150	25.5	0.013	0.13	–	0.33(2.5)
H_2O	25	75.3	0.46	0.65	6.49	–
$LiNO_3$	600	111.3	0.149	1.48	8.49	20,9(25)
$NaNO_3$	600	155.3	0.134	1.25	10.50	30(30)
KNO_3	600	123.4	0.260	2.35	10.98	20(25)
$RbNO_3$	600	–	0.300	2.75	15.96	–
$HgBr_2$	450	102.1	0.579	5.20	–	20(5)
$HgCl_2$	450	102.1	0.509	4.59	–	19(5)
HgJ_2	450	102.1	0.670	5.20	–	11.4(4)
$BiCl_3$	300	143.5	0.365	2.48	–	6(3)

*The numbers in brackets are values of pressure in kbar.

do not exceed 0.07 % of C_p. This implies that at pressure up to 10 bar the following equality can be assumed to high accuracy:

$$C_U = C_P. \tag{2.8}$$

Equality (2.8) suggests that at the conditions mentioned above the internal energy of a substance U can be assumed equal to the quantity of enthalpy H listed in reference books on thermodynamic properties of substances [20-30]:

$$U = H. \tag{2.9}$$

Henceforth, equality (2.9) will be used at normal pressure.

It is significant that at pressure higher than 100 bar the quantity of αPV becomes noticeable in comparison with the quantity of C_p. So at 600 °C the quantity of αPV for KNO_3 makes 0.2, 1.9, and 8.9 % at pressures of 100, 1000, and 10000 bar respectively, while at a pressure of 25000 bars it increases up to 16.2 % of C_p. Therefore at high pressures it is necessary to use equation (2.2) to define the internal energy of a substance. At normal pressure equation (2.7) becomes simpler:

$$dU = C_p dT. \tag{2.10}$$

If we integrate this equation over a temperature interval from 0K up to some temperature T which should not exceed a temperature of the first phase transition, we will find at low pressures:

$$U_T - U_0 = \int_0^T C_p dT. \tag{2.11}$$

A change in internal energy being counted off from an unknown quantity U_0, it is convenient to assume that $U = U_T - U_0 = 0$ at $T \to 0$ and to refer to the quantity U as to the internal energy of a substance. From above it follows that at pressure up to 10 bar it is possible to assume:

$$U = \int_0^T C_p dT. \tag{2.12}$$

In order to determine the internal energy of a substance to be at temperature higher than that of the first phase transition, energy of phase transition $\Delta_{trs}U$ has to be added to equation (2.12). Thus, the latter should be re-written in the following way:

$$U = \int_0^{T_1} C_p dT + \Delta_{trs}U + \int_{T_1}^{T} C_p dT. \tag{2.13}$$

The quantity C_p in the second integral covers temperatures above the temperature of first phase transition T_1. Energy of other phase transitions is taken into account by analogy.

Fig. 2.1 shows the temperature dependence of the internal energy for nickel in temperature range from 0 up to 2400K taking into account the data on the temperature (1726K) and energy (17.62 kJ/mol) of melting. The data in Fig. 2.1 are normalized to $U_0 = G_0 = 0$ at $T \to 0$.

2.2. The second law of thermodynamics at constant pressure and temperature dependence of entropy and free energy of substance

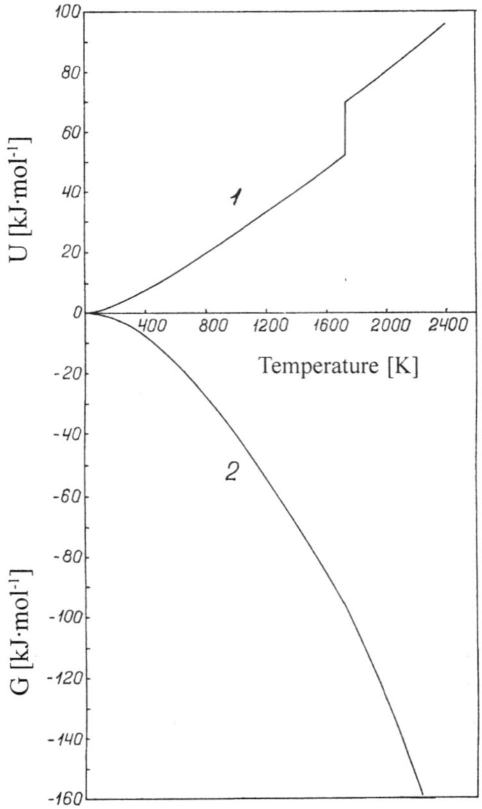

Fig. 2.1. Temperature dependence of internal (curve 1) and free (curve 2) energies for nickel.

As is known from experience, a selected substance can participate in various physical (evaporation, transition into a solution) and chemical (decomposition, formation of chemical compounds) processes besides an interchange of energy with other substances. These processes are controlled with another level of internal energy G that coincides with a level of U at $T = 0$K, it is located below U_0 (see Fig. 2.1, curve 2). The quantity G is referred to as the free energy; it is related to the internal energy U of a substance by the equation

$$U = G + TS, \tag{2.14}$$

where the quantity S is called the entropy of a substance. Entropy of a substance is the thermal coefficient of free energy:

$$S = -\left(\frac{\partial G}{\partial T}\right)_P, \quad (2.15)$$

which is determined from experience. For this purpose, let us differentiate equation (2.14) and use (2.15). As a result we have:

$$dU = TdS. \quad (2.16)$$

Equations (2.14) and (2.16) are those of the second law of thermodynamics in the integral and differential forms.

In order to derive the equation for temperature dependence of entropy of a substance, we use equations (2.16) and (2.7). Thus we arrive at the equation

$$\left(\frac{\partial S}{\partial T}\right)_P = \frac{1}{T}(C_P - \alpha PV), \quad (2.17)$$

which can be written at normal pressure as:

$$dS = \frac{C_P}{T}dT. \quad (2.18)$$

By integrating this equation over temperature from 0K up to some temperature T, which does not exceed temperature of the first phase transition, we obtain:

$$S - S_0 = \int_0^T \frac{C_P}{T}dT. \quad (2.19)$$

Entropy of any substance is equal to zero at $T \to 0$K according to the third law of thermodynamics. Consequently: $S_0 = 0$.

When phase transition of a substance from one condensed state to another takes place (polymorphic transformation and melting) with the energy of transition being $\Delta_{trs}U$, the entropy of the substance rises by the quantity:

$$\Delta_{trs}S = \frac{\Delta_{trs}U}{T_{trs}}. \quad (2.20)$$

By taking into consideration the entropy of the first phase transition we arrive at the following equation for calculating entropy of a substance at temperatures above that of the first phase transition (T_1):

$$S = \int_0^{T_1} \frac{C_P}{T}dT + \Delta_{trs}S + \int_{T_1}^{T} \frac{C_P}{T}dT. \quad (2.21)$$

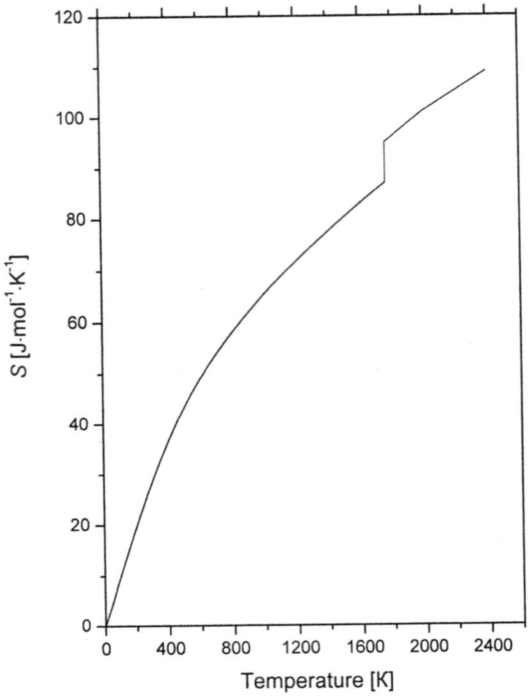

Fig. 2.2. Temperature dependence of entropy for nickel.

Other phase transitions are taken into account by analogy. Fig. 2.2 shows the entropy of nickel as the function of temperature in the temperature interval from 0 up to 2400K, with the fact that the entropy of melting for nickel makes 10.21 $J \cdot mol^{-1} \cdot K^{-1}$ at temperature 1726K taken into account.

The most convenient way to calculate the temperature dependence of the free energy of a substance is to use the equation:

$$G = U - TS \qquad (2.22)$$

with the values of U and S known. The equation holds for phase transitions as well. It follows in view of equation (2.20), that the free energy of the phase transitions listed above is equal to zero:

$$\Delta_{trs} G = 0. \qquad (2.23)$$

The free energy of nickel calculated in the range of temperatures from 0 up to 2300K using equation (2.22), with equation (2.23) taken into consideration is presented in Fig. 2.1 (curve 2). In the melting point of nickel at 1726K the slope of the curve G, T increases due to the entropy of melting.

2.3. The third law of thermodynamics at constant pressure

Experience suggests that the energy of any substance being extracted its temperature decreases. As a result, the substance finally becomes unable to release energy, and therefore acquires to the least level of the internal energy U_0. Experience also shows that at the same time, the temperature tends to $T = -273{,}15$ °C, i.e. to zero temperature (0K).

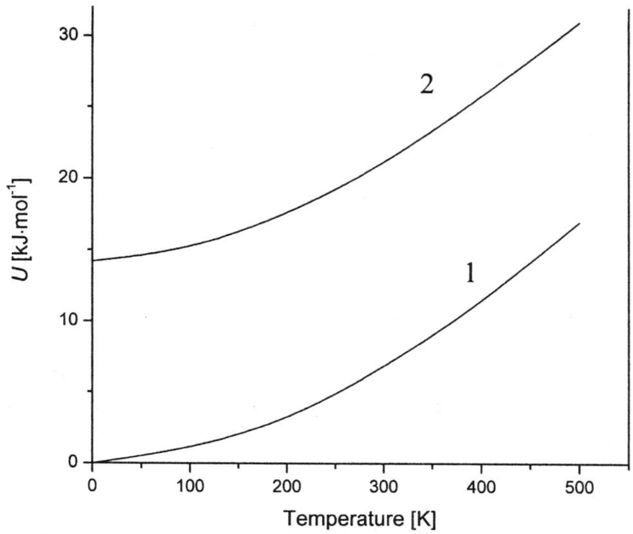

Fig. 2.3. Change in the internal energy of crystalline (cristobalite, curve 1) and glassy (curve 2) silica in the temperature interval from 0 up to 500K.

The level of internal energy U_0 of the same substance depends on its structure. For example, if the melt of a substance has not crystallized on cooling and has turned to amorphous state (glass), the quantity U_0 for this state of the substance is higher, than that for initial crystal:

$(U_0)_{am} > (U_0)_{cr}$.

The variation in the internal energy of crystalline (cristobalite) and glassy silica with the temperature ranging from 0 to 500K is presented in Fig. 2.3.

At construction of this figure $(U_0)_{cr} = 0$ has been assumed for crystalline silica. In this case the difference $(U_0)_{am} - (U_0)_{cr}$ makes 14.2 kJ·mol^{-1}. This value is not much different from the melting energy for cristobalite which equals 9.58 kJ·mol^{-1} at 2001K.

The first and second laws of thermodynamics are stated as the laws of change in the internal energy of a substance. Hence, we shall formulate the third law of thermodynamics by the same principle. As noted above, any substance close to zero temperature loses its ability to release energy. This phenomenon can be expressed mathematically in the following way:

$$dU_{T\to 0} = 0. \qquad (2.24)$$

The third law of thermodynamics, as well as the previous ones, does not depend on structure of substance. Referring to Fig. 2.3 it is to observe that the slope of the U,T curve for crystalline and amorphous silica tends to zero nearby the zero temperature. This fact is clearly illustrated by Fig. 1.4 which shows the change in the internal energy of glassy silica at temperatures up to 1K.

In other words, the third law of thermodynamics does not depend on an absolute level of energy U_0 which a substance possesses at $T \to 0$K. The level of energy U_0 can be increased not only by turning the substance into amorphous state (see Fig.2.3), but by a buildup of pressure. Therefore the third law of thermodynamics can be written in general terms at any constant pressure:

$$dU_{P=const,\ T\to 0} = 0. \tag{2.25}$$

That is to say that the free energy of a substance nearby zero temperature ceases to change at any constant pressure according to the third law of thermodynamics:

$$dG_{P=const,\ T\to 0} = 0. \tag{2.26}$$

In view of (2.15) from equation (2.26) follows:

$$S_{P=const,\ T\to 0} = 0. \tag{2.27}$$

Thus, entropy of any substance, irrespective of its structure and pressure, is equal to zero if the temperature is close to absolute zero. The preceding applies to amorphous substances, glasses, and solid solutions as well.

2.4. The first law of thermodynamics at constant temperature and pressure dependence of the internal energy of substance

If a condensed substance is compressed in a thermostat, it is necessary to remove an amount of heat dq from the thermostat in order to maintain its temperature constant. Experience suggests that dq is less than the work of compression dw done on the substance ($dw > dq$). The thing is that the internal energy of the substance has increased by an amount dU^P at compression.

Mathematically, the energy balance can be expressed as the equation:

$$dU^P = dw - dq. \tag{2.28}$$

In this equation dU^P denotes the variation of the internal energy with the pressure. Equation (2.28) is that of the first law of thermodynamics for a case when the internal energy of a substance increases due to work of compression. The minus sign before dq implies that heat is transferred to the surroundings.

If applying the law of conservation of energy to the energy in the form of work of compression dw ($dw = dU^P + dq$) we arrive at the conclusion that the amount of heat dq transferred to the surroundings does not exceed the work of compression dw. Hence follows that: $dU^P > 0$. Thus, it is possible to conclude, relying on the first law of thermodynamics at constant temperature:

The internal energy of a condensed substance increases with the growth of pressure at constant temperature.

Mathematically, the above can be expressed as follows:

$$\left(\frac{\partial U}{\partial P}\right)_T > 0. \tag{2.29}$$

As is known, the internal energy of ideal gas does not depend on pressure.

Let us consider the dependence of internal energy on pressure.

In classical thermodynamics using the concept of enthalpy:

$$H = U + PV \tag{2.30}$$

and the relations:

$$dq_{rev} = dH = TdS_{cl}, \tag{2.31}$$

$$TdS_{cl} = dU + PdV, \tag{2.32}$$

$$\left(\frac{\partial S_{cl}}{\partial P}\right)_T = -\left(\frac{\partial V}{\partial T}\right)_P, \tag{2.33}$$

where S_{cl} is the entropy in classical thermodynamics, q_{rev} is the heat of a reversible process, the following equation for the pressure dependence of internal energy at constant temperature has been derived:

$$\left(\frac{\partial U}{\partial P}\right)_T = V(k_T P - \alpha T). \tag{2.34}$$

In this equation, k_T and α are the isothermal compressibility and cubic expansion coefficients for a substance, V, P, and T are the molar volume, pressure, and thermodynamic temperature.

It follows from the calculation of $V(k_T P - \alpha T)$ for many substances [8,9] that the derivative $(\partial U/\partial P)_T$ is negative. This clearly shows that the internal energy of a substance decreases with the growth of pressure, according to classical thermodynamics. That is possible only provided that when being isothermally compressed the substance releases more heat dq to the environment than the work of compression done dw. From the law of conservation of energy it follows that such interchange of energy is impossible, as the substance cannot transfer more heat to the environment than the work of compression done.

The fundamentally incorrect result following from equation (2.34) is due to the fact that classical thermodynamics postulates the following:

– the mathematical product PV is considered as energy on level with the internal energy of substance,
– entropy S_{cl} is considered a property of heat, whereas substance is known to possess not heat, but only internal energy.

The above has been taken into account when deriving the pressure dependence of internal energy of substance presented below.

Equation (2.1) can be rewritten as follows:

$$dU^P = -PdV - dq. \tag{2.35}$$

The minus sign before the product PdV is evident from $dV < 0$. Having divided both parts of the equation by dP we obtain at constant temperature:

$$\left(\frac{\partial U^P}{\partial P}\right)_T = -P\left(\frac{\partial V}{\partial P}\right)_T - \left(\frac{\partial q}{\partial P}\right)_T. \tag{2.36}$$

Taking into account

$$\left(\frac{\partial V}{\partial P}\right)_T = -k_T V, \tag{2.37}$$

after denoting:

$$\beta_q = \left(\frac{\partial q}{\partial P}\right)_T \tag{2.38}$$

we have the desired equation

$$\left(\frac{\partial U^P}{\partial P}\right)_T = D = k_T PV - \beta_q. \tag{2.39}$$

In equation (2.39) D is the pressure coefficient of internal energy, k_T is the isothermal compressibility, β_q – is the pressure coefficient of the heat released by a substance at isothermal compression.

$D = 0$ at $P \to 0$. Consequently:

$$\left(\frac{\partial U^P}{\partial P}\right)_{T,P\to 0} = 0, \tag{2.40}$$

i.e., at low pressure the internal energy of substance does not depend on pressure. At $P >> 0$ we have $D > 0$. Consequently, the internal energy of substance increases with the growth of pressure.

At present the pressure coefficient of heat β_q in equation (2.39) for pressure coefficient D of the internal energy of substance is unknown for all substances under investigation. Therefore thermodynamic properties of individual substances at high pressures can be determined only under the assumption that $\beta_q = 0$. Then from equation (2.39) follows

$$D = k_T PV. \tag{2.41}$$

As the isothermal compressibility k_T in this equation is defined by the equation:

$$k_T = -\frac{1}{V}\left(\frac{\partial V}{\partial P}\right)_T, \tag{2.42}$$

it is necessary to have the experimental data on the pressure dependence of molar volume V of substance in order to define the pressure coefficient D of the internal energy of any substance at various pressure.

With $\beta_q = 0$, from equation (2.39) it follows that:

$$dU^P = DdP. \tag{2.43}$$

Upon integrating the equation with respect to the pressure from the lower limit P_0, which is usually equal to standard pressure $P^0 = 1$ bar at which $U^P = 0$, up to pressure P we obtain:

$$U^P = \int_{P^0}^{P} DdP. \tag{2.44}$$

Thus, the internal energy of a substance at the chosen temperature and at some pressure P is equal to:

$$U = U^0 + \int_{P^0}^{P} DdP, \tag{2.45}$$

where U^0 is the internal energy of the substance at pressure $P^0 = 1$ bar and temperature T.

The standard values of internal energy U^0 at $P^0 = 1$ bar for many substances are listed in reference books [20 – 25]. Having constructed the curve D,P using the curve V,P and equations (2.41) and (2.42) equation (2.44) can be integrated. Thus the internal energy U can be found for a selected substance at various pressure through the use of (2.45).

The curve U^P,P for aluminum at 298K is shown in Fig. 2.4 (curve 1). In this case $U^0 = 4.58$ kJ·mol^{-1}. As pressure grows up to 200 kbar the internal energy of aluminum increases up to 19.6 kJ·mol^{-1}, i.e. more than four times.

2.5. The second law of thermodynamics at constant temperature and pressure dependence of the free energy of substance

In the equation of the second law of thermodynamics at constant pressure entropy S

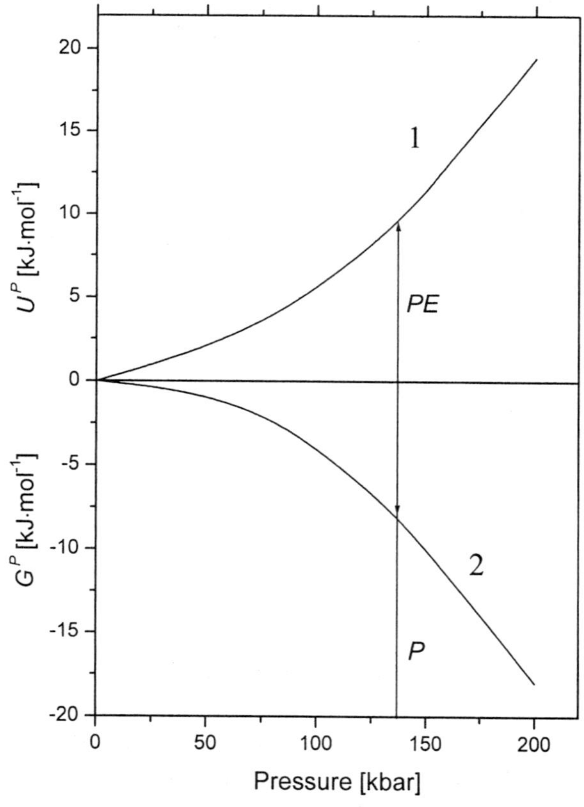

Fig. 2.4. Change in the internal (curve 1) and free (curve 2) energies for aluminum at 298K with the buildup of pressure from normal up to 200 kbar ($T = 298$K).

$$dU = TdS \tag{2.46}$$

is related to the free energy by the equation:

$$-dG = SdT. \tag{2.47}$$

At constant temperature an increment in the internal energy dU^P and a decrease of free energy $-dG^P$ at the buildup of pressure by dP are defined by the following equations:

$$dU^P = PdE, \tag{2.48}$$

$$-dG^P = EdP, \tag{2.49}$$

where E is the pressure coefficient of the free energy of a substance.

Equation (2.48) is the expression of the second law of thermodynamics for an increase in the internal energy of a substance due to isothermal compression. Quantities U^P, G^P, and E are related to each other by the equation

$$U^P = G^P + PE, \tag{2.50}$$

from which the defining equation for pressure follows:

$$P = \frac{U^P - G^P}{E}. \tag{2.51}$$

Equation (2.50) is illustrated by Fig. 2.4.

The pressure coefficient of free energy E in equations (2.48) and (2.49) is equal to the slope ratio of the curve G^P,P at the given pressure:

$$-\left(\frac{\partial G^P}{\partial P}\right)_T = E. \tag{2.52}$$

$E = 0$ at $P \to 0$ and from equation (2.52) follows:

$$\left(\frac{\partial G^P}{\partial P}\right)_{T,P\to 0} = 0, \tag{2.53}$$

i.e., at low pressure the free energy of a substance does not depend on pressure.

When $P>>0$ we have $E>0$. As a results, from equation (2.52) it follows that the free energy of a substance decreases with growth of pressure.

The magnitude of pressure coefficient of free energy E also depends on pressure. The equation of this dependence can be derived from the following equalities:

$$dU^P = DdP, \tag{2.54}$$

$$dU^P = PdE. \tag{2.55}$$

Equating these equations we have:

$$\left(\frac{\partial E}{\partial P}\right)_T = \frac{D}{P}. \tag{2.56}$$

From equation (2.56) it follows that the pressure coefficient of the free energy of a substance increases with the growth of pressure. As noted above, $D = 0$ at $P \to 0$, and consequently:

$$\left(\frac{\partial E}{\partial P}\right)_{T,P\to 0} = 0, \tag{2.57}$$

i.e. at low pressure the pressure coefficient of free energy does not depend on pressure.

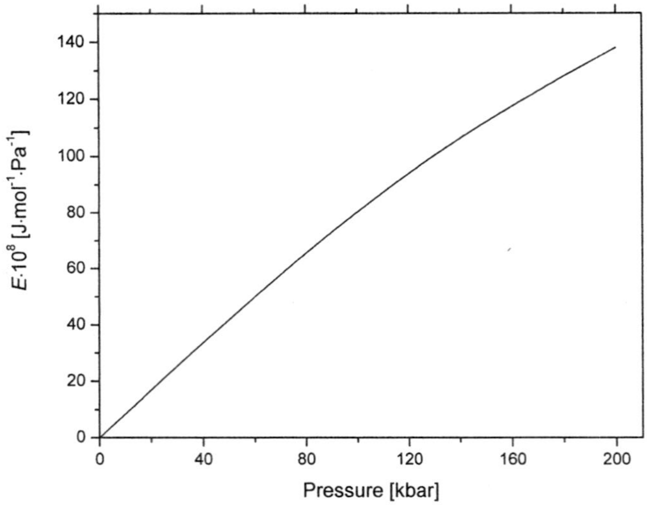

Fig. 2.5. Pressure dependence of the pressure coefficient of free energy for aluminum at 298K.

Upon integrating equation (2.56) between the limits P_0 at which $E = 0$ and some value P we arrive at

$$E = \int_{P_0}^{P} \frac{D}{P} dP. \tag{2.58}$$

Having constructed the curve *D/P*, *P* by using an available dependence *D,P*, we integrate equation (2.58) and thus arrive at the *E,P* curve which is used to define the pressure dependence of free energy *G* of a substance. The *E,P* curve for aluminum at 298K is shown in Fig. 2.5.

In order to define the curve *G, P* the following equation should be integrated:

$$-dG^P = EdP\ . \tag{2.59}$$

Inasmuch as $G^P = 0$ at the lower limit of integration $P_0 = P^0 = 1$ bar, from equation (2.59) follows:

$$G^P = -\int_{P_0}^{P} EdP. \tag{2.60}$$

The equation is integrated with the use of the *E,P* curve found.

The pressure dependence of free energy G^P for aluminum at 298K is pictured in Fig. 2.4 (curve 2).

The quantity G^P can also be determined with equation (2.50).

2.6. The third law of thermodynamics at constant temperature

At constant temperature and low pressure the internal energy of a substance does not depend on pressure:

$$(dU^P)_{P\to 0, T=const}=0. \tag{2.61}$$

The equation is an expression of the third law of thermodynamics at constant temperature. From equation (2.61) and the second law of thermodynamics at constant temperature (see equation (2.50)) follows:

$$(dG^P)_{P\to 0, T=const}=0. \tag{2.62}$$

As $dG^P = -EdP$, we have at $P \to 0$:

$$(E)_{P\to 0, T=const}=0. \tag{2.63}$$

Thus, the slope ratio of curves U^P,P and G^P,P at $P \to 0$ is equal to zero, according to the third law of thermodynamics at constant temperature. This fact is illustrated by Fig. 2.4.

3

LAWS OF THERMODYNAMICS AT ISOBARIC–ISOTHERMAL INCREASE IN THE SURFACE AREA OF SUBSTANCE

When studying the action of temperature and pressure on the internal and free energies of a macrosize substance influence of the surface on the properties mentioned is neglected. This approach is quite justified, for in this case mole fraction of atoms or molecules in the surface of a substance is of the order of 10^{-7}. The energy of particles in the surface of a nanodispersed substance with the particles in size from 1 up to 100 nm should also be taken into account because the fraction of particles in the surface can be higher than 0.5.

In the chapter the laws of thermodynamics are discussed at isobaric-isothermal increase in the surface area of a substance.

3.1. Surface of substance and interfacial area

The concept of the surface of a substance is often used in thermodynamics. Consequently, we have to be specific about its meaning. Strictly speaking, a surface can be referred only when a solid or a liquid (i.e. the condensed substance) borders on vacuum. However, the situation is extremely seldom realized in experiment. Therefore, referring to a surface makes sense when the condensed substance is in equilibrium with its vapor or a gas which is not dissolved in the substance. Henceforth, the concept of surface of a substance is implicit in the above definition. In all other cases it would be reasonable to speak about phase interface or interfacial area.

Thus we assume that the surface area of nanoparticles can be calculated based upon their geometry if the minimum size of nanoparticles is greater than one nanometer. The equations for calculating molar surface areas of a nanodispersed substance for three cases are listed below.

Case 1. A mole of the substance, which makes full sphere of radius r_0 is divided into z spherical particles of radius r. Molar surface area of z spherical particles equals:

$$A = az = 4\pi r^2 z, \tag{3.1}$$

where a is the surface area of a spherical particle.

Assuming that the density of substance is independent of the size of a particle we have:

$$z = \frac{V}{\nu} = \frac{r_0^3}{r^3}. \tag{3.2}$$

Here V and v are the molar volume and the volume of a nanoparticle.

Using the last two equations and taking into account that $V = (4/3)\pi r_0^3$ we find:

$$A = \frac{3V}{r}. \tag{3.3}$$

Case 2. Strings with radius of r and length $100r$ are produced from a cylinder of volume V, height $2r_0$, and radius r_0. The number of strings z is equal to $r_0^3 / 50r^3$. If $r << r_0$ for this case we obtain the equation

$$A = \frac{2V}{r}. \tag{3.4}$$

Case 3. The plates of dimensions $d_0 \cdot d_0 \cdot d$, where d is the thickness of a plate, are produced from a cube of volume V and d_0 on an edge. As the number of such plates equals d_0/d then at $d << d_0$ in this case we find the equation

$$A = \frac{2V}{d}. \tag{3.5}$$

As appears, equations (3.3), (3.4), and (3.5) differ in multiplier factors 3 and 2. Factor 3 suggests that the dispersity of substance is "three-dimentional", while factor 2 indicates to its "two-dimentionality".

3.2. The first law of thermodynamics. Specific surface energy of substance

It is known from experience that a certain work needs to be done to increase the surface area of a liquid. This means that the internal energy of atoms or molecules in the surface of a liquid is higher, than that inside the liquid. The aforesaid applies to any condensed substance. Let us denote the excess internal energy of a mole of substance distributed in its surface as compared with the internal energy U of bulk substance ($z = 1$) as U^S. A certain work dw has to be done to increase U^S by an amount dU^S and the surface area of a substance by dA. From the first law of thermodynamics follows that the work dw is spent not only for an increase in the internal energy of a substance dU^S, but also on the heat dq released to the environment.

The energy balance between dw, dU^0, and dq is a mathematical expression of the first law of thermodynamics for the isobaric-isothermal increase in the surface area of a substance:

$$dw = dU^S + dq. \tag{3.6}$$

The magnitude dU^S in the equation above is proportional to an increment in the surface area dA:

$$dU^S = \sigma\, dA. \tag{3.7}$$

The proportionality constant σ in the equation above shows what the increase is in the excess internal energy U^S in the surface of one mole of substances due to the increase in the surface area by $1 m^2$. Therefore quantity σ according to the defining equation:

$$\sigma = \left(\frac{\partial U^S}{\partial A}\right)_{P,T} \tag{3.8}$$

will be named the specific surface energy, omitting, for simplicity sake, “internal”. The specific surface energy has dimensionality $[J \cdot m^{-2}]$. Let us divide both parts of equation (3.6) by dA at constant pressure and temperature:

$$\left(\frac{\partial w}{\partial A}\right)_{P,T} = \left(\frac{\partial U^S}{\partial A}\right)_{P,T} + \left(\frac{\partial q}{\partial A}\right)_{P,T}. \tag{3.9}$$

Introducing the designations:

$$\sigma_W = \left(\frac{\partial w}{\partial A}\right)_{P,T}, \tag{3.10}$$

$$\sigma_q = \left(\frac{\partial q}{\partial A}\right)_{P,T}, \tag{3.11}$$

we rewrite equation (3.9) as follows:

$$\sigma_W = \sigma + \sigma_q \quad . \tag{3.12}$$

In the equation above, σ_W is called the surface tension, according to classical thermodynamics. Firstly, such a definition does not agree with dimensionality σ_W $[J \cdot m^{-2}]$. Secondly, it is the work for an increase in the surface that is characterized by σ_W, and therefore σ_W is not a property of a substance by contrast with its specific surface energy σ.

The quantity σ_W can be measured experimentally but only for liquids. Taking into account that the magnitude σ_q can be neglected in equation (3.12) we have the equality:

$$\sigma_W = \sigma, \tag{3.13}$$

which will be used below.

Up to the moment specific surface energy has been measured only for liquids. Some of the data are listed in Table 3.1.

As can be seen from the table, the specific surface energies of various liquids and melts vary over a wide interval of values: from 0.022 J·m^{-2} for ethanol at 293K up to 2.44 J·m^{-2} for liquid tungsten at 3683K.

Table 3.1. Specific surface energies of some liquids and melts [11, 29, 31].

Substance, formula	Temperature [K]	Specific surface energy σ [J·m^{-2}]
Water H_2O	293	0.0727
Ethanol C_2H_6O	293	0.0220
Methanol CH_4O	293	0.0226
Acetone C_3H_6O	293	0.0237
Glycerin $C_3H_8O_3$	293	0.0594
Mercury Hg	293	0.4862
Sodium Chloride NaCl	1073	0.1140
Boron oxide B_2O_3	773	0.0669
Silicon oxide SiO_2	2063	0.380
Silicon Si	1685	0.578
Silver Ag	1273	0.922
Gold Au	1338	1.170
Platinum Pt	2073	1.800
Lead Pb	623	0.442
Bismuth Bi	553	0.372
Tungsten W	3683	2.44
Copper Cu	1373	1.320
Iron Fe	1823	1.780
Cobalt Co	1773	1.790
Nickel Ni	1773	1.810

3.3. Specific interfacial energy

The defining equation (3.8) of the specific surface energy σ applies to a certain condensed substance or a phase. If the given phase 1 contacts with a phase 2 (1 and 2 are the condensed phases), at the boundary of phase 1 with phase 2 the specific surface energy of phase 1 is equal to σ_{12}. Thus the specific surface energy of phase 2 at the boundary with phase 1 is equal to σ_{21}. Magnitudes σ_{12} and σ_{21} differ from σ_1 and σ_2 at the boundary with air or with its vapor. However it is only the specific interfacial energy σ_{LL} that can be measured at the boundary between two liquids. This is due to the fact that the energy measured at increase in the surface area of the contact between two liquids (liquid / liquid) is related to the surfaces of both liquids at the boundary of their contact, i.e. to the interfacial surface.

Specific interfacial energies of mercury with various liquids are presented in Table 3.2.

If comparing the data of the given table with the specific surface energy of mercury $\sigma_{Hg} = 0.486$ J·m^{-2} at 20 °C and the specific surface energies of liquids contacting with mercury one may notice that the specific interfacial energy σ_{HgL} is less than σ_{Hg}, but greater than σ_L.

Values of specific interfacial energy at the boundary between water and some organic liquids as well as those for water and the liquids are given in Table 3.3.

For the water/liquid system as well as for the mercury/liquid system the following inequalities are valid: $\sigma_L < \sigma_{H_2O\,L} < \sigma_{H_2O}$.

Table 3.2. Specific interfacial energies σ_{HgL} at the boundary of mercury with various liquids and specific surface energies of the liquids σ_L at 20 °C [J·m^{-2}] (Hg – mercury, L – a liquid)

Liquid, formula	σ_{HgL}	σ_L
Water H_2O	0.375	0.0727
Aniline C_3H_6N	0.341	0.0433
Acetone C_3H_6O	0.390	0.0240
Benzene C_6H_6	0.357	0.0286
Hexane C_6H_{14}	0.378	0.0184
Nitrobenzene $C_6H_5NO_2$	0.350	0.0439
Octane C_8H_{18}	0.375	0.0216
Carbon disulfide CS_2	0.339	0.0322
Toluene C_7H_8	0.359	0.0285
Tetrachloromethane CCl_4	0.362	0.0270

Table 3.3. Specific interfacial energies $\sigma_{H_2O\,L}$ at the boundary of water with organic liquids and specific surface energies of the liquids at 20 °C [J·m^{-2}].

Liquid, formula	$\sigma_{H_2O\,L}$	$\sigma_{L.}$
Water	–	0.0727
Benzene C_6H_6	0.0337	0.0284
Toluene C_7H_8	0.0357	0.0285
Chlorbenzene C_6H_5Cl	0.0381	0.0332
Tetrachloromethane CCl_4	0.0438	0.0270
Carbon bisulfide CS_2	0.0483	0.0322
Cyclohexane C_6H_{12}	0.0510	0.0247
Hexane C_6H_{14}	0.0512	0.0195
Decane $C_{10}H_{22}$	0.0512	0.0239

3.4. On the defining equation for the specific surface energy of substance in classical thermodynamics

As the defining equation for the specific surface energy (3.8) is derived with the help of the first law of thermodynamics, it would be interesting to compare it with the defining equation for the specific surface energy:

$$\sigma = \left(\frac{\partial G}{\partial A}\right)_{P,T}, \qquad (3.14)$$

assumed in classical thermodynamics [32] (at $G = G^G$, see section 1.2.1). As appears, the equation above does not coincide with (3.8). This is due to the fact that according to classical thermodynamics the work for increasing the surface area of liquid is considered to be spent on the increase in the free energy:

$$dw = dG, \qquad (3.15)$$

which decreases in actual fact. Equation (3.15) contradicts the first law of thermodynamics that connects work dw with internal energy dU not free energy dG. Therefore equation (3.14) is not used in the book.

3.5. Dependence of the surface energy of substance on its surface area

Molar internal energy of a substance is equal to U at the initial surface area A_0. As molar surface area A increases, the internal energy of a substance is augmented due to excess U^S, i.e. at the cost of the surface energy. Therefore, the total molar internal energy U^d of a dispersed substance equals:

$$U^d = U + U^S. \tag{3.16}$$

Upon integrating equation (3.8) at the constant value σ we find:

$$U^S = \sigma\,(A - A_0)\ . \tag{3.17}$$

With regard to the equation above from (3.16) follows:

$$U^d = U + \sigma(A - A_0). \tag{3.18}$$

For a nanodispersed substance, $A >> A_0$. Consequently, equations (3.17) and (3.18) can be rewritten in the following way:

$$U^S = \sigma A, \tag{3.19}$$

$$U^d = U + \sigma A. \tag{3.20}$$

Thus, the dependence U^S, A shows up as a straight line with the slope of σ. When $A = A_0$ it gives $U^S = 0$ (see Fig.3.1, straight line 1). The dependence U^d, A is also linear with a slope equal to σ. At $A = A_0$ we have $U^d = U$ (see Fig.3.1, straight line 2). Equations (3.17) – (3.20) are valid when $A < A_1$, where A_1 is the surface area at the minimum size of nanoparticles equal to one nanometer.

When a nanodispersed substance consists of spherical particles of radius r, from equations (3.3) and (3.19) follows:

$$U^S = \frac{3\sigma V}{r}. \tag{3.21}$$

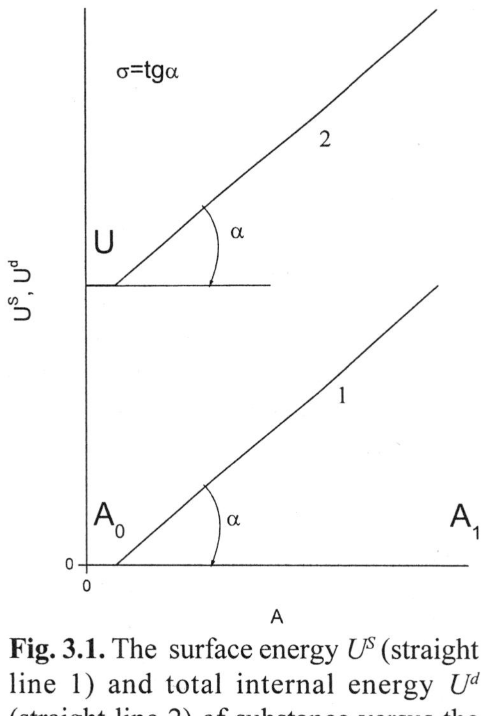

Fig. 3.1. The surface energy U^S (straight line 1) and total internal energy U^d (straight line 2) of substance versus the surface area A. The slope of straight lines U^S,A and U^d,A is equal to the specific surface energy of substance. U is the internal energy of bulk substance at the set temperature. A_0 and A_1 – see the explanation in the text.

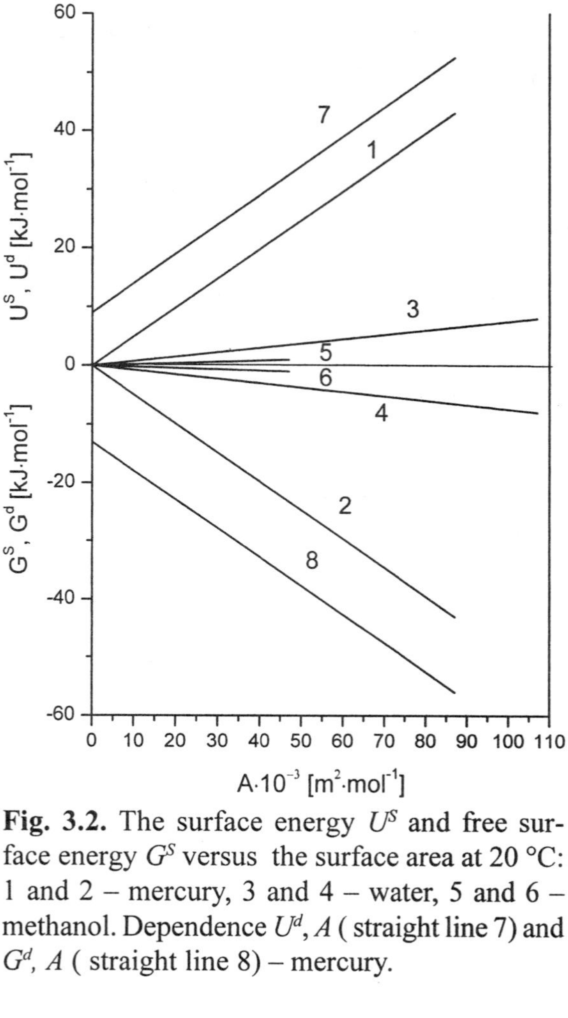

Fig. 3.2. The surface energy U^S and free surface energy G^S versus the surface area at 20 °C: 1 and 2 – mercury, 3 and 4 – water, 5 and 6 – methanol. Dependence U^d, A (straight line 7) and G^d, A (straight line 8) – mercury.

If a nanodispersed substance consists of strings of radius r, in view of equation (3.4) we have

$$U^S = \frac{2\sigma V}{r}. \tag{3.22}$$

If a nanodispersed substance consists of plates with dimensions $d_0 \cdot d_0 \cdot d$, with regard to equation (3.5) we find:

$$U^S = \frac{2\sigma V}{d}. \tag{3.23}$$

As is evident, the surface energy of a substance is inversely proportional to the radius of a sphere and a string or to the thickness of a plate.

Let us apply equation (3.17) to the drops of mercury, water, and methanol at 20 °C. The data required for the calculation of magnitudes A, U^S, and U^d are listed below. The magnitude A_l has been calculated by equation (3.3) at $r = 0.5 \cdot 10^{-9}$ m.

	Hg	H_2O	CH_4O
V [m³·mol⁻¹]	$14.821 \cdot 10^{-6}$	$18.048 \cdot 10^{-6}$	$40.429 \cdot 10^{-6}$
r_0 [m]	0.01525	0.01625	0.0117
σ [J·m⁻²]	0.486	0.0727	0.0226
U [J·mol⁻¹]	9215	–	–
A_1 [m²·mol⁻¹]	89146	107859	242574
A_0 [m²·mol⁻¹]	0.00292	0.00332	0.00744

The data obtained are presented in Table 3.4 and in Fig. 3.2.

The dependence U^S, A for the drops Hg, H_2O, and CH_4O and the dependence U^d, A for Hg are straight lines (see Fig.3.2, straight lines 1, 3, 5, 7). The increase in the internal energy of mercury, water, and methanol, due to an increase in

Table 3.4. Values A, U^S, and G^S at various diameters of drops of mercury, water, and methanol at 20 °C. For mercury: $U = 9215$ J·mol^{-1}, $G = -12896$ J·mol^{-1}.

d[m]	A[m^2·mol^{-1}]	U^S[J·mol^{-1}]	G^S[J·mol^{-1}]
		Hg	
$d_0 = 0.0305$	0.00292	0	0
$8.915 \cdot 10^{-5}$	1.0	0.486	-0.486
10^{-7}	891.5	433	-433
10^{-8}	8915	4332	-4332
10^{-9}	89146	43325	-43325
		H_2O	
$d_0 = 0.0325$	0.00332	0	0
$1.078 \cdot 10^{-4}$	1.0	0.073	-0.073
10^{-7}	1078.6	78.4	-78.4
10^{-8}	10786	784	-784
10^{-9}	107860	7841	-7841
		CH_4O	
$d_0 = 0.0426$	0.00744	0	0
$2.426 \cdot 10^{-4}$	1.0	0.0226	-0.0226
10^{-7}	2425.7	54.8	-54.8
10^{-8}	24257	548	-548
10^{-9}	242574	5480	-5480

molar surface area from A_0 up to A_l comes to 43325, 7841, and 5840 J·mol^{-1} respectively. Thus the internal energy of mercury increases from 9215 up to 43325 J·mol^{-1} i.e. by factor 4.7 (see straight line 7, Fig. 3.2).

3.6. The second and third laws of thermodynamics. Specific free surface energy of substance

As is well known (section 3.2), the U^S, A dependence is linear, with the slope equal to specific surface energy:

$$\left(\frac{\partial U^S}{\partial A}\right)_{P,T} = \sigma . \tag{3.24}$$

Therefore the dependence of free surface energy G^S on the surface area of a substance is also linear with the slope equal to specific free surface energy of a substance f:

$$f = -\left(\frac{\partial G^S}{\partial A}\right)_{P,T}. \tag{3.25}$$

Let us find the quantity f with the help of the third law of thermodynamics. Before writing the equation of the third law of thermodynamics at constant temperature and pressure, let us now review the content of the third law of thermodynamics at P = const and A = const. In this case the third law consists in the following:
firstly, $U_0 = G_0 = 0$ at $T \to 0$;
secondly, the slopes of curves U,T and G,T are identical and equal to zero at $T \to 0$:

$$\left(\frac{\partial U}{\partial T}\right)_{P,A,T\to 0} = -\left(\frac{\partial G}{\partial T}\right)_{P,A,T\to 0} = 0. \tag{3.26}$$

In other words, $C_U = 0$ and $S = 0$ at $T \to 0$.

As applied to isobaric-isothermal increase in the surface area of a substance the law can be written by analogy with the preceding and taking into account that $\sigma_0 = \sigma$ at $A \to A_0$:
firstly, $U^S = G^S = 0$ at $A \to A_0$,
secondly, the slopes of straight lines U^S,A and G^S,A are identical and opposite in sign at $A \to A_0$:

$$\left(\frac{\partial U^S}{\partial A}\right)_{P,T,A\to A_0} = -\left(\frac{\partial G^S}{\partial A}\right)_{P,T,A\to A_0} = \sigma. \tag{3.27}$$

A comparison between the equality above and equations (3.24), (3.25) shows that $f = \sigma$ and equation (3.25) must be written as follows:

$$-\left(\frac{\partial G^S}{\partial A}\right)_{P,T} = \sigma. \tag{3.28}$$

Thus, the curves U^S,A and G^S,A are symmetric about the axis A, and $U_0^S = G_0^S = 0$ at A_0. Then the second law of thermodynamics in the integrated form can be expressed by means of the equation

$$U^S - G^S = 2\sigma(A - A_0). \tag{3.29}$$

3.7. Dependence of the free surface energy of substance on its surface area

Upon integrating the equation

$$dG^S = -\sigma dA \tag{3.30}$$

over the interval from $G_0^S = 0$ at $A = A_0$ up to G^S at some value A, we have

$$dG^S = -\sigma(A - A_0), \tag{3.31}$$

which may be simplified at $A >> A_0$:

$$G^S = -\sigma A. \tag{3.32}$$

The dependences G^S, A for the drops of mercury, water, and methanol are presented in Fig. 3.2 (curves 2,4, and 6) and in Table 3.4.

If a nanodispersed substance consists of spherical particles, equation (3.32) becomes:

$$G^S = -\frac{3\sigma V}{r}. \tag{3.33}$$

If a nanodispersed substance consists of strings or plates (see section 3.1) equation (3.32) can be written:

$$G^S = -\frac{2\sigma V}{r}, \tag{3.34}$$

$$G^S = -\frac{2\sigma V}{d}. \tag{3.35}$$

The total free energy G^d of a nanodispersed substance is equal to

$$G^d = G + G^S, \tag{3.36}$$

where G is the free energy of the substance at $A = A_0$. The dependence G^d, A for the drops of mercury is shown in Fig. 3.2 (straight line 8).

3.8. The thermodynamic equation for defining the specific surface energy of solid at melting temperature

The surface area of a nanodispersed solid changes from A_S up to A_L at melting. At the same time the free surface energies of a solid G_S^S and a liquid G_L^S are equal

$$G_S^S = -\sigma_S A_S, \tag{3.37}$$

$$G_L^S = -\sigma_L A_L. \tag{3.38}$$

In these equations σ_S and σ_L are the specific surface energies of a solid and a liquid at melting temperature.

If a spherical particle of a solid is melted the formed liquid holds the spherical form, and at melting temperature the equality is observed:

$$G_S^d = G_L^d. \tag{3.39}$$

As

$$G_S^d = G_S + G_S^S, \tag{3.40}$$

$$G_S^d = G_L + G_L^S, \tag{3.41}$$

from equation (3.39) with the account of G_S=G_L the equality follows:

$$G_S^S = G_L^S. \tag{3.42}$$

In equations (3.40)–(3.42) G_S,G_L are the free energies of bulk substance, G_S^d, G_L^d and G_S^S, G_L^S are the total free energies and free surface energies of nanodispersed substance in solid and liquid states.

With the help of equations (3.42), (3.37), and (3.38) we find the equality

$$\sigma_S A_S = \sigma_L A_L, \tag{3.43}$$

from which follows

$$\sigma_S = \sigma_L \frac{A_L}{A_S}. \tag{3.44}$$

If a mole of substance consists of z spherical particles with the surface area a and with volume v, the molar area A and the molar volume V of a substance are equal

$$A = za, \tag{3.45}$$

$$V = zv. \tag{3.46}$$

Table 3.5. Specific surface energies of some solid metals at melting temperatures (T_m) calculated by equation (3.49).

Metal	T_m [K]	$V_L \cdot 10^{-6}$ [$m^3 \cdot mol^{-1}$]	$V_S \cdot 10^{-6}$ [$m^3 \cdot mol^{-1}$]	σ_L [$J \cdot m^{-2}$]	σ_S [$J \cdot m^{-2}$]
Li	453.6	13.557	13.357	0.406	0.410
Na	370.9	24.800	24.195	0.208	0.211
K	336.8	47.220	46.046	0.102	0.104
Rb	312	57.593	56.192	0.086	0.087
Mg	923	15.312	14.851	0.580	0.592
Al	933	11.360	10.717	0.860	0.894
Ga	302.9	11.437	11.803	0.715	0.700
Si	1683	10.928	12.088	0.725	0.678
Pb	600	19.404	18.748	0.470	0.481
Bi	544.4	20.794	21.515	0.390	0.381
Sb	903.7	18.731	18.910	0.350	0.348
Zn	692.7	9.983	9.581	0.806	0.828
Cd	594.2	14.016	13.387	0.642	0.662
Hg	234.3	14.655	14.132	0.498	0.510
Cu	1356	8.003	7.688	1.350	1.387
Ag	1234	11.599	11.174	0.910	0.933
Au	1338	11.398	10.838	1.170	1.210

As the surface area a of a sphere and its volume v are interrelated by the proportion

$$a = \sqrt[3]{36\pi v^2}, \tag{3.47}$$

from equations (3.45) and (3.46) we find:

$$\frac{A_L}{A_S} = \left(\frac{V_L}{V_S}\right)^{2/3}. \tag{3.48}$$

Substituting the equation above into (3.44) gives the equation for defining the specific surface energy of a solid at melting temperature:

$$\sigma_S = \sigma_L \left(\frac{V_L}{V_S}\right)^{2/3}. \tag{3.49}$$

In equations (3.48) and (3.49) V_L and V_S are the molar volumes of a substance in liquid and solid states respectively. Equation (3.49) is valid for bulk substance ($z = 1$).

Let us apply equation (3.49) to gold. In this case at melting temperature of 1338K $V_L = 11.398 \cdot 10^{-6}$, $V_S = 10.838 \cdot 10^{-6}$ $m^3 \cdot mol^{-1}$ and $\sigma_L = 1.170$ $J \cdot m^{-2}$. With the help of equation (3.49) we find that $\sigma_S = 1.210$ $J \cdot m^{-2}$. As is evident, the magnitude σ_S exceeds σ_L only by 3.4 %. The molar volume of silicon decreases at melting from $V_S = 11.977 \cdot 10^{-6}$ down to $V_L = 10.928 \cdot 10^{-6}$ $m^3 \cdot mol^{-1}$ at 1683K. Consequently, the specific surface energy of silicon at fusion decreases from 0.725 down to 0.682 $J \cdot m^{-2}$, i.e. by 6.2 %.

The specific surface energies of some solid metals calculated by equation (3.49) at melting temperatures are listed in Table 3.5.

If $V_L > V_S$ then $\sigma_S > \sigma_L$, and if $V_L < V_S$ $\sigma_S < \sigma_L$ according to equation (3.49).

4

TEMPERATURE AND PRESSURE DEPENDENCE OF THE THERMODYNAMIC PROPERTIES OF NANODISPERSED SUBSTANCE

As a rule, the specific surface energy of a substance decreases if both temperature and pressure increase. Therefore, the influence of the surface area on the thermodynamic properties of a substance decreases as the temperature and pressure grow. This problem has been treated at length in the chapter below.

4.1. Temperature dependence of the specific surface energy of liquid

Specific surface energy σ, defined by the equations

$$\sigma = \left(\frac{\partial U^S}{\partial A}\right)_{P,T}, \tag{4.1}$$

$$\sigma = -\left(\frac{\partial G^S}{\partial A}\right)_{P,T} \tag{4.2}$$

depends on temperature. From experience it is known that, as a rule, σ decreases as temperature increases, i.e.:

$$\left(\frac{\partial \sigma}{\partial T}\right)_{P,A} < 0. \tag{4.3}$$

Some melts are exceptions to this rule, by way of example, zinc or boron oxide (B_2O_3) melts. In addition, the specific surface energy decreases linearly for a wide range of liquids and melts as temperature increases:

$$\sigma = \sigma_0 + \left(\frac{\partial \sigma}{\partial T}\right)_{P,A}(T - T_0), \tag{4.4}$$

where σ_0 is the specific surface energy at temperature T_0 which is less than temperature T, $(\partial\sigma/\partial T)_{P,A}$ is the temperature coefficient of σ, determined experimentally with dimensionality [J·m^{-2}·K^{-1}]. Fig. 4.1 shows the temperature dependence of specific surface energy for liquid gold in the interval 1338 – 1600K, which is described by the equation:

$$\sigma = 1.170 - 0.10\cdot 10^{-3}(T - T_m), \tag{4.5}$$

where the melting temperature T_m equals 1338K.

The temperature dependence of σ for mercury over an interval of 0 – 100 °C is described by the equation

$$\sigma = 0.4906 - 0.2049 \cdot 10^{-3}(T - 273) \qquad (4.6)$$

and is presented in Fig. 4.2. Noticeable deviations of dependence σ,*T* from linearity are observed for water (see Fig.4.3).

The temperature dependence of the surface energy of liquid metals Na, K, Rb, and Cs is described by a power series:

$$\sigma = \sigma_0 - a(T - T_m) - b(T - T_m)^2 + c(T - T_m)^3 ,$$

where σ_0 is the specific surface energy of the above-listed metals near melting temperature, while *a*, *b*, and *c* are coefficients.

The temperature dependence of the specific surface energy for some liquids is given in Table 4.1. The specific surface energies of the liquids listed in the table decrease as temperature increases.

Table 4.1. Specific surface energies of some inorganic and organic liquids [29, 31] at temperatures 10 – 100 °C.

Substance, formula	σ [mJ·m^{-2}] at temperature [^{0}C]				
	10	25	50	75	100
Water H_2O	74.23	71.99	67.94	63.57	58.91
Mercury Hg	488.55	485.48	480.36	475.23	470.11
Bromine Br	43.68	40.95	36.40		
sulfuric acid H_2SO_4	52.92	52.39	51.70		
tetrachloromethane CCl_4	26.43	23.37	20.31	17.25	
carbon bisulfide CS_2	33.81	31.58	27.87		
methanol CH_4O	23.23	22.07	20.14		
acetic acid $C_2H_4O_2$	28.8	27.3	24.8	22.13	
ethanol C_2H_6O	23.22	21.97	19.89		
acetone C_3H_6O	23.46	20.66			
methylacetate $C_3H_6O_2$	26.66	24.73	21.51		
1- propanol C_3H_8O	24.48	23.32	21.38	19.43	
2- propanol C_3H_8O	22.11	20.93	18.96	16.98	
glycerin $C_3H_8O_3$	59.2	58.0			
diethyl ether $C_4H_{10}O$	19.4	16.4	13.5		
benzene C_6H_6	30.24	28.22	25.0	21.77	
toluene C_7H_8	27.71	27.93	24.96	21.98	19.01

4.2. Temperature dependence of the specific surface energy of solid

As is well known (see section 3.8), the specific surface energy of solid at melting temperature can be determined by the equation

$$\sigma_S = \sigma_L \left(\frac{V_L}{V_S} \right)^{2/3}, \tag{4.7}$$

where σ_S, σ_L, and V_S, V_L are the specific surface energies and the molar volumes of a substance in solid and liquid states.

The melting temperature decreasing as a result of reduction in the diameter of spherical particles of a solid, equation (4.7) is also valid for nanodispersed substance at temperatures below the melting temperature of bulk substance.

The molar volume of a liquid V_L at temperatures below the melting temperature can be estimated by extrapolating experimental curve V_L,T for the liquid.

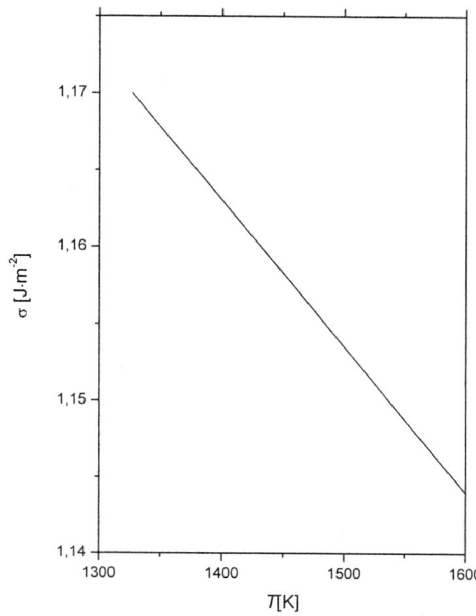

Fig. 4.1. Temperature dependence of the specific surface energy of liquid gold (the melting temperature equals 1338K).

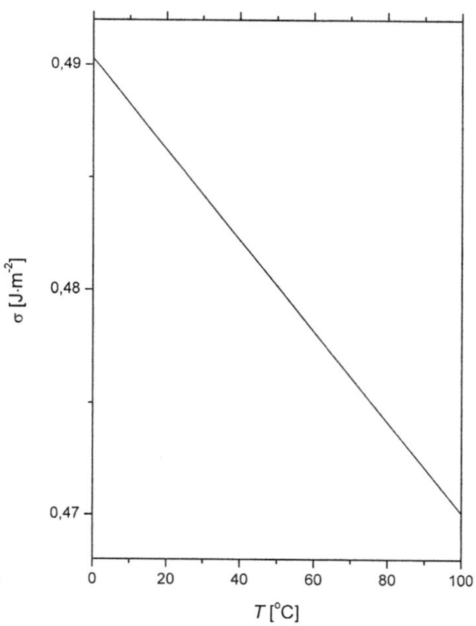

Fig. 4.2. Temperature dependence of the specific surface energy of mercury.

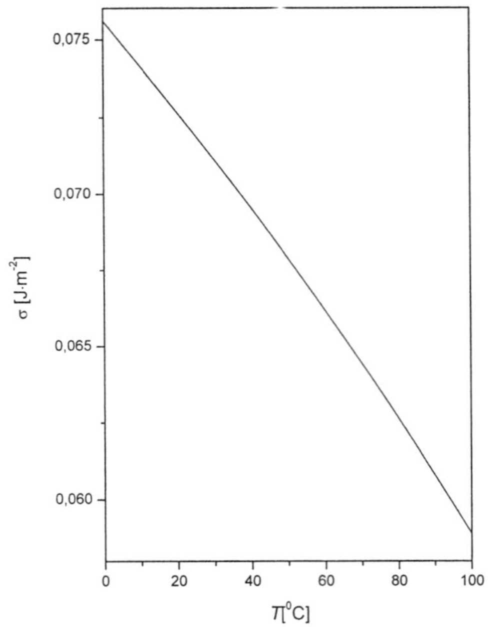

Fig. 4.3. Temperature dependence of the specific surface energy of water.

For example, the molar volume of liquid metals in the certain temperature interval is described by the equation

$$V_L = V_L^0 + b(T - T_m), \qquad (4.8)$$

where V_L^0 is the molar volume of a liquid at $T = T_m$. Therefore, the following equation can be used to estimate the molar volume of a liquid at $T < T_m$:

$$V_L = V_L^0 + b(T_m - T). \qquad (4.9)$$

If the V_S,T curve for a solid is not known, but its molar volumes $V_{S,\,298}$ and V_S^0 at 298K and melting temperature are, the equation of line may be used to estimate V_S:

$$V_S = V_S^0 + c(T - 298), \qquad (4.10)$$

where the coefficient c is estimated by the equation

$$c = \frac{V_S^0 - V_{S,298}}{T_m - 298}. \qquad (4.11)$$

We have for gold $V_L^0 = 11.398\cdot10^{-6}$ and $V_S^0 = 10.838\cdot10^{-6}$ $m^3\cdot mol^{-1}$, $b = 7.58\cdot10^{-10}$, $c = 5.99\cdot10^{-10}$ $m^3\cdot mol^{-1}\cdot K^{-1}$. Therefore, in this case

$$V_L = 11.398\cdot10^{-6} - 7.58\cdot10^{-10}\,(1338 - T), \qquad (4.12)$$

$$V_S = 10.838\cdot10^{-6} - 5.99\cdot10^{-10}\,(1338 - T). \qquad (4.13)$$

The specific surface energy of liquid gold is described by equation [11]

$$\sigma_L = 1.170 - 1.0\cdot10^{-4}\,(T - 1338). \qquad (4.14)$$

In order to estimate a magnitude of σ_L at T <1338K, let us assume that linear dependence (4.14) is kept. Thus at T <1338K we have:

$$\sigma_L = 1.170 + 1.0\cdot10^{-4}\,(1338 - T). \qquad (4.15)$$

The melting temperatures of gold T at different radii of drops are presented in Fig. 5.2.

Table 4.2. Specific surface energies of solid gold at temperatures 298 – 1338K calculated by equation (4.7).

T_m [K]	$V_L \cdot 10^{-6}$ [$m^3 \cdot mol^{-1}$]	$V_S \cdot 10^{-6}$ [$m^3 \cdot mol^{-1}$]	σ_L [$J \cdot m^{-2}$]	σ_S [$J \cdot m^{-2}$]
298	10.609	10.215	1.274	1.307
400	10.687	10.276	1.264	1.297
500	10.763	10.336	1.254	1.288
600	10.838	10.396	1.244	1.279
700	10.914	10.456	1.234	1.270
800	10.990	10.516	1.224	1.260
900	11.066	10.575	1.214	1.251
1000	11.142	10.635	1.204	1.242
1100	11.218	10.695	1.194	1.233
1200	11.293	10.755	1.184	1.223
1250	11.331	10.785	1.179	1.218
1338	11.398	10.838	1.170	1.210

The values σ_S calculated for gold at various temperatures by equation (4.7) are listed in Table 4.2. The necessary values V_L, V_S, and σ_L have been estimated by means of equations (4.12), (4.13), and (4.15).

The dependence σ_S,T for gold is described by the equation

$$\sigma_S = 1.210 + 0.936 \cdot 10^{-4} (1338 - T). \qquad (4.16)$$

A comparison between equations (4.15) and (4.16) suggests that as temperature decreases, the specific surface energy of the solid increases more slowly than that of the liquid.

4.3. Temperature dependence of the internal and free energies of nanodispersed substance

The internal U^d and free G^d energies of a nanodispersed substance (see chapter 3) are defined with the help of the equations

$$U^d = U + U^S\,, \qquad (4.17)$$

$$G^d = G + G^S\,, \qquad (4.18)$$

where the surface energy U^S and surface free energy G^S are equal to:

$$U^S = \sigma A\,, \qquad (4.19)$$

$$G^S = -\sigma A\,. \qquad (4.20)$$

With regard to (4.19) and (4.20), from equations (4.17) and (4.18) follows:

$$U^d = U + \sigma A\,, \qquad (4.21)$$

$$G^d = G - \sigma A. \qquad (4.22)$$

In these equations U is the internal and G is the free energy of bulk substance. The temperature dependence of them has been discussed in chapter 2. The values U and G for the solid and the liquid states of many substances have been listed in reference books [20–30].

By way of illustration, let us consider the plotting of the U^d,T and G^d,T curves at a constant surface area A. If the drops of mercury have drop diameters d of 100, 10, and 1nm at 20 °C, the surface areas A are equal to 891.5, 8915, and 89146 $m^2 \cdot mol^{-1}$ respectively. Below we shall generate the temperature dependence of U^d and G^d for the listed values A for mercury in the temperature range from 0 up to 100 °C.

The temperature dependence of U and G has been determined with the help of reference data [23, 27] taking into account that $U_0 = G_0 = 0$ at $T \to 0$. The value of σ was determined by equation (4.6). The calculated values of U^S, U^d and G^S, G^d for mercury are presented below and in Fig. 4.4 and 4.5.

T [^{0}C]	U [$J \cdot mol^{-1}$]	U^S [$J \cdot mol^{-1}$]	U^d [$J \cdot mol^{-1}$]
		$A = 891.5\ m^2 \cdot mol^{-1}$	
0	8690	437.4	9127
25	9343	432.8	9776
50	10050	428.2	10478
75	10720	423.7	11144
100	11420	419.1	11839
		$A = 8915\ m^2 \cdot mol^{-1}$	
0	8690	4374	13064
25	9343	4328	13671
50	10050	4282	14332
75	10720	4237	14957
100	11420	4191	15611

		$A = 89146$ m^2·mol^{-1}	
0	8690	43740	52430
25	9343	43280	52623
50	10050	42820	52870
75	10720	423370	53090
100	11420	41910	53330

T [°C]	G [J·mol^{-1}]	G^S [J·mol^{-1}]	G^d [J·mol^{-1}]
		$A = 891.5$ m^2·mol^{-1}	
0	-11250	-437.4	-11687
25	-13250	-432.8	-13683
50	-15050	-428.2	-15478
75	-17150	-423.7	-17574
100	-19220	-419.1	-19639
		$A = 8915$ m^2·mol^{-1}	
0	-11250	-4374	-15624
25	-13250	-4328	-17578
50	-15050	-4282	-19332
75	-17150	-4237	-21387
100	-19220	-4191	-23411
		$A = 89146$ m^2·mol^{-1}	
0	-11250	-43740	-54990
25	-13250	-43280	-56490
50	-15050	-42820	-57870
75	-17150	-42370	-59520
100	-19220	-41910	-61130

The data above imply that U^S decreases with increasing temperature, while G^S increases (see Fig.4.4, curves 2 and 4, Fig. 4.5, curves 2 and 4). The point is that the specific surface energy of mercury decreases as a result of an increase in temperature (see Fig.4.2). If temperature increases, so does the magnitude U^d increases, while G^d decreases (see Fig.4.4, curves 3 and 5, Fig. 4.5, curves 3 and 5). At the same time, the slopes of the U^d,T and G^d,T curves at A = const are less than the ones of the U,T and G,T curves (see Fig.4.4, curve 1, Fig. 4.5, curve 1).

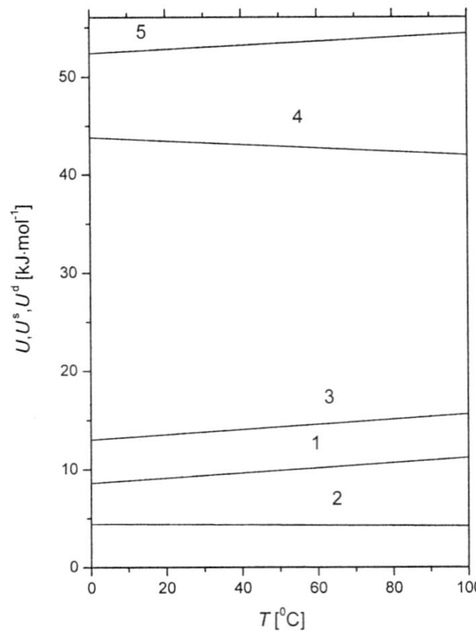

Fig. 4.4. Temperature dependence of U (curve 1), U^S, and U^d for mercury at A = 8915 m^2·mol^{-1} (curve 2 – U^S,T and curve 3 – U^d,T) and A = 89146 m^2·mol^{-1} (curve 4 – U^S,T and curve 5 – U^d,T).

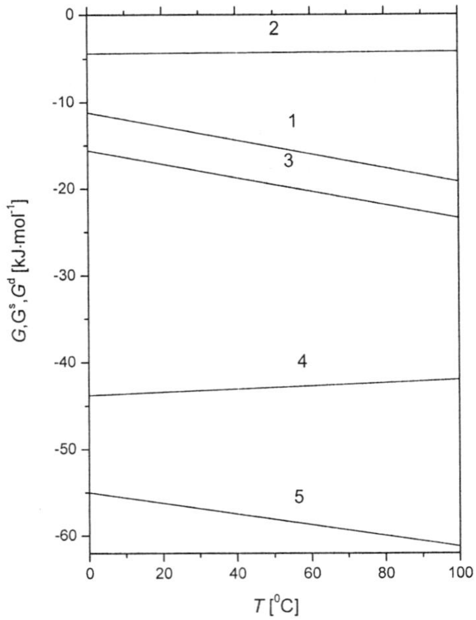

Fig. 4.5. Temperature dependence of G (curve 1), G^S, and G^d for mercury at A = 8915 m^2·mol^{-1} (curve 2 – G^S,T and curve 3 – G^d,T) and A = 89146 m^2·mol^{-1} (curve 4 – G^S,T and curve 5 – G^d,T).

4.4. Entropy of nanodispersed substance and its temperature dependence

Dividing both parts of equations (4.18) and (4.22) by dT at A = const and P = const we have:

$$\left(\frac{\partial G^d}{\partial T}\right)_{P,A} = \left(\frac{\partial G}{\partial T}\right)_{P,A} + \left(\frac{\partial G^S}{\partial T}\right)_{P,A}, \tag{4.23}$$

$$\left(\frac{\partial G^d}{\partial T}\right)_{P,A} = \left(\frac{\partial G}{\partial T}\right)_{P,A} - \left(\frac{\partial \sigma}{\partial T}\right)_{P,A} \cdot A. \tag{4.24}$$

In these equations

$$\left(\frac{\partial G^d}{\partial T}\right)_{P,A} = -S^d, \tag{4.25}$$

$$\left(\frac{\partial G}{\partial T}\right)_{P,A} = -S, \tag{4.26}$$

$$\left(\frac{\partial G^S}{\partial T}\right)_{P,A} = -S^S. \tag{4.27}$$

In addition,

$$S^d = S + S^S = S + s^S A\ . \tag{4.28}$$

With regard to (4.25) – (4.28) from equation (4.24) follows:

$$s^S = \left(\frac{\partial \sigma}{\partial T}\right)_{P,A}. \tag{4.29}$$

In equations (4.25) – (4.29) S^d and S are the entropies of a nanodispersed and bulk substance, S^S and s^S are surface and specific surface entropies of a substance. The quantity s^S has dimensionality [J·m^{-2}·K^{-1}].

From the preceding follows that the temperature dependence of the specific surface entropy s^S can be found using experimental curve σ,T. The temperature dependence of surface entropy S^S of a substance at a set A will be obtained using both the s^S,T curve and equation (4.28). Further the temperature dependence of entropy S^d for nanodispersed substance is determined by equation (4.28) and the known temperature dependence of entropy S for bulk substance.

In the previous section the internal energy U^d and free energy G^d of mercury have been calculated at 20 °C and the values A equal 891.5, 8915 and 89146 m^2·mol^{-1}. As a further example the entropy S^d for mercury has been calculated at the specified values A in an interval of temperatures from 0 up to 100 °C. The temperature dependence of S has been determined using reference data [23], the value of σ has been determined by equation (4.6). The data obtained are shown below.

T [°C]	S[J·mol^{-1}·K^{-1}]	S^S [J·mol^{-1}·K^{-1}]	S^d [J·mol^{-1}·K^{-1}]
		A = 891.5 m^2·mol^{-1}	
0	73.04	-0.1827	72.86
25	75.81	–”–	75.63
50	77.71	–”–	77.53

75	80.08	–»–	79.89
100	82.14	–»–	81.96
		$A = 8915$ m²·mol⁻¹	
0	73.04	-1.827	71.21
25	75.81	–»–	73.98
50	77.71	–»–	75.88
75	80.08	–»–	78.25
100	82.14	–»–	80.31
		$A = 89146$ m²·mol⁻¹	
0	73.04	-18.27	54.77
25	75.81	–»–	57.54
50	77.71	–»–	59.44
75	80.08	–»–	61.81
100	82.14	–»–	63.87

As the specific surface entropy of mercury s^S does not depend on temperature, nor does S^S at a constant value of A. The quantity S^d increases with increasing temperature, as S grows with increased temperature. The entropy of nanodispersed substance S^d decreases with an increase in the surface area, as the specific surface entropy $s^S < 0$. For example, for mercury from equation (4.6) follows $s^S = -0.2049 \cdot 10^{-3}$ J·m⁻²·K⁻¹. With regard to equation (4.28) for mercury we find the equation $S^d = S - 0.2049 \cdot 10^{-3} A$.

4.5. Pressure dependence of the internal and free energies of nanodispersed substance

The equations of pressure dependence of internal and free energies of nanodispersed substance can be derived by dividing both parts of the following equations:

$$U^d = U + \sigma A, \quad (4.30)$$

$$G^d = G - \sigma A \quad (4.31)$$

by dP at T = const and A = const. Then we have:

$$\left(\frac{\partial U^d}{\partial P}\right)_{T,A} = \left(\frac{\partial U}{\partial P}\right)_{T,A} + \left(\frac{\partial \sigma}{\partial P}\right)_{T,A} \cdot A, \quad (4.32)$$

$$\left(\frac{\partial G^d}{\partial P}\right)_{T,A} = \left(\frac{\partial G}{\partial P}\right)_{T,A} - \left(\frac{\partial \sigma}{\partial P}\right)_{T,A} \cdot A. \quad (4.33)$$

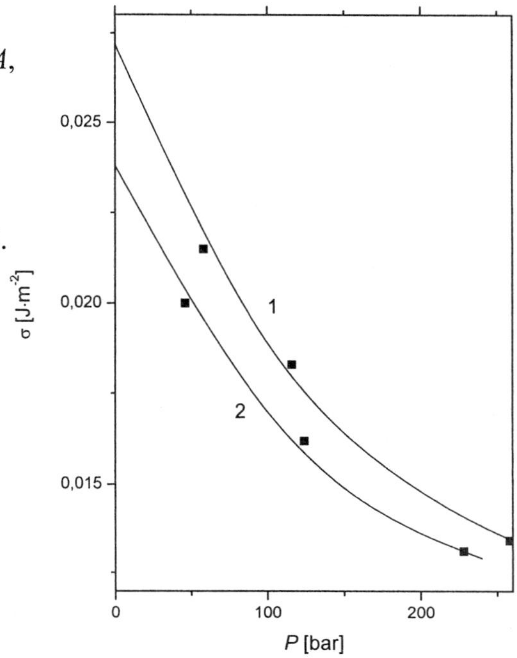

Fig. 4.6. Pressure dependence of the specific surface energy of oil at temperatures 20 °C (curve 1) and 60 °C (curve 2).

With regard to the defining equations

$$D^d = \left(\frac{\partial U^d}{\partial P}\right)_{T,A}, \quad (4.34)$$

$$D = \left(\frac{\partial U}{\partial P}\right)_{T,A}, \quad (4.35)$$

$$E^d = -\left(\frac{\partial G^d}{\partial P}\right)_{T,A}, \quad (4.36)$$

$$E = -\left(\frac{\partial G}{\partial P}\right)_{T,A} \quad (4.37)$$

we have:

$$D^d = D + \left(\frac{\partial \sigma}{\partial P}\right)_{T,A} \cdot A, \quad (4.38)$$

$$E^d = E + \left(\frac{\partial \sigma}{\partial P}\right)_{T,A} \cdot A. \quad (4.39)$$

In equations (4.34) – (4.39) D,E and D^d,E^d are the pressure coefficients of internal and free energy of bulk and nanodispersed substances respectively.

The magnitudes D and E can be calculated basing the experimental pressure dependence of molar volume V of a substance (see chapter 2). An experimental pressure dependence of σ has to be available in order to determine magnitudes U^d and G^d. The pressure dependence of specific surface energy for oil (Tuymaza, Russia) in an atmosphere of nitrogen at temperatures 20 and 60 °C are presented below.

20 °C				
Pressure [bar]	1	58.8	115.5	258.4
σ [$J \cdot m^{-2}$]	0.0272	0.0215	0.0114	0.0134
60 °C				
Pressure [bar]	1	45.6	123.6	232
σ [$J \cdot m^{-2}$]	0.0238	0.0200	0.0162	0.0131

The data are given in graphical form in Fig. 4.6. It suggests that the pressure coefficient of specific superficial energy of oil decreases with the growth of pressure. The pressure coefficients of the specific surface energy of oil at 20 and 60 °C are tabulated below.

temperature [°C]	$-(\partial\sigma/\partial P)_{T,A} \cdot 10^{10}$ [$J \cdot m^{-2} \cdot Pa^{-1}$] at pressures [bar]				
	1	50	100	150	200
20	11.1	7.85	5.00	3.97	2.83
60	9.08	7.26	4.64	3.33	2.22

As appears, the pressure coefficient of specific surface energy decreases by a factor of 3.9 at 20 °C and by the factor 4.1 at 60 °C as the pressure grows from 1 up to 200 bar.

5

EQUILIBRIUM BETWEEN THE PHASES OF NANODISPERSED SUBSTANCE

Equilibrium curves for two phases and triple points of a substance are determined not only by temperature and pressure but by the surface area as well. In other words, the *P,T* diagram of phase equilibrium for one substance is a section of the *P,T,A* diagram at the constant surface area *A* of a substance. The influence of the surface area of a substance on phase equilibrium for one substance is discussed in the current chapter.

5.1. Phase diagram for macrophases of one substance and equations for equilibrium lines of two phases

The general phase diagram for one substance is presented in Fig. 5.1. The data for nitrogen (continuous lines) are taken as a basis. Dashed lines are drawn arbitrarily. The mentioned diagram has been discussed in detail in [9]. Therefore we shall briefly treat the phase diagram for one substance shown in Fig. 5.1.

What makes this diagram different from one accepted in classical thermodynamics is that a normal liquid disappears and instead of it there is the second form of a liquid which is has been named fluid (region F on Fig. 5.1).

The lines of the diagram in fig. 5.1 refer to equilibriums between the following phases:
fa – the solid and the vapor (sublimation),
ab – the liquid and the vapor (evaporation),
bc - the fluid and the vapor (evaporation of the fluid),
bd – the liquid and the fluid,
ad – the solid and the liquid (melting),
dc – the solid and the fluid (melting of the fluid),
ck – the solid and the vapor (sublimation).

The consideration of these phase equilibriums suggests that they may be divided into two groups:
first, equilibrium between two condensed phases,
second, equilibrium between the condensed phase and its vapor.

For the first group of phase equilibriums, a change in temperature dT_{tr} of the transition of one condensed phase 1 into other phase 2 and an increase in external pressure by dP are interrelated by the equation [8,9]:

$$\frac{dT_{tr}}{dP} = -\frac{E_2 - E_1}{S_2 - S_1} = -\frac{\Delta_{tr}E}{\Delta_{tr}S}, \tag{5.1}$$

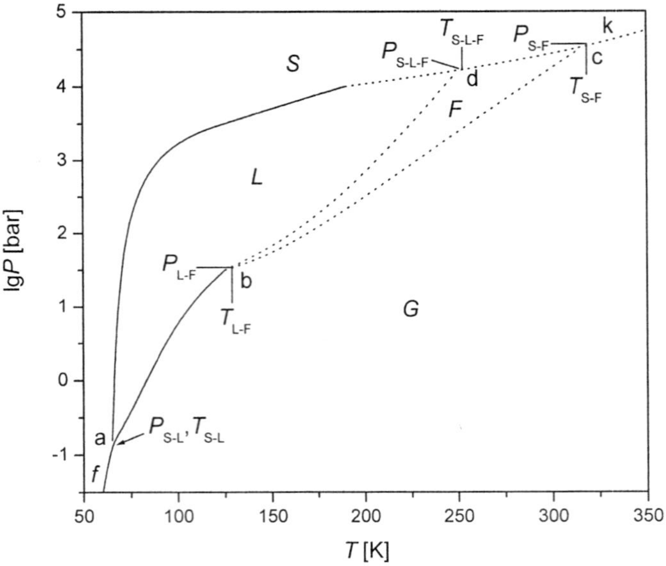

Fig. 5.1. The general phase diagram for one substance. The data for nitrogen (continuous lines) are taken as a basis. Dashed lines are drawn arbitrarily. Lines of equilibrium between two phases: *fa, ab, bc,* and *ck* concern to equilibrium of the vapor with the solid, liquid, fluid, and solid respectively; *ad, bd, dc* refer to the equilibrium of the liquid with the solid, fluid, and the fluid with the solid. The fields of existence of one phase: *S* – the solid, *L* – the liquid, *F* – the fluid, *G* – the gas. Points *a, b, c,* and *d* – the triple points (see [9]).

where E_1, E_2 are the pressure coefficients of free energy of phases 1 and 2, S_1, S_2 are the entropies of phases 1 and 2, $\Delta_{tr}E$ and $\Delta_{tr}S$ are the changes in the pressure coefficient of free energy and entropy at phase transition.

For the second group of phase transitions an increase in the saturated vapor pressure at evaporation and sublimation by an amount dP_S and an increase in temperature by dT are interrelated by the equations [8, 9]:

$$\frac{dP_S}{dT} = \frac{\Delta_{vap}S}{\Delta_{vap}V - \Delta_{vap}E}, \tag{5.2}$$

$$\frac{dP_S}{dT} = \frac{\Delta_{sub}S}{\Delta_{sub}V - \Delta_{sub}E}, \tag{5.3}$$

where $\Delta_{vap}S$ and $\Delta_{sub}S$, $\Delta_{vap}V$ and $\Delta_{sub}V$, $\Delta_{vap}E$ and $\Delta_{sub}E$ are the change in entropy, volume, and pressure coefficient of free energy at evaporation and sublimation.

The lines of phase equilibrium shown in Fig. 5.1 are described by equations (5.1), (5.2), and (5.3).

5.2. Dependence of saturated vapor pressure on the surface area for nanodispersed substance

As far as a condensed substance with a constant surface area A is concerned, it is well known that the relationship between an increase in saturated vapor pressure by dP_S and a decrease of free energy of vapor formation $d\Delta_{cd\text{-}g}G$ owing to increase in temperature by dT is expressed by the equation [8,9]:

$$-d\Delta_{cd-g}G = \left[V(g) - V(cd)\right]dP_S, \tag{5.4}$$

where $V(g)$ and $V(cd)$ are the molar volumes of vapor and a condensed substance, $\Delta_{cd\text{-}g}G = G(g) - G(cd)$ is the free energy of vapor formation.

If pressure is low its influence on both the free energy of vapor formation $\Delta_{cd\text{-}g}G$ and the volume of a condensed substance $V(cd)$ may be neglected. Then [8, 9] equation (5.4) becomes:

$$-d\Delta_{cd\text{-}g}G = RTd\ln P_S\,. \tag{5.5}$$

Upon integrating the given equation within the range from $P^0 = 1$ bar, when $\Delta_{cd\text{-}g}G = 0$, to a certain value P_S we obtain the well-known equation:

$$\ln\frac{P_S}{P^0} = -\frac{\Delta_{cd-g}G}{RT}. \tag{5.6}$$

If the surface area of a substance is increased (for example, by crushing it) at constant temperature, saturated vapor pressure is known to build up to P_S at an amount A in comparison with P_S^* at $A = A_0$. As in this case increase in the pressure of saturated vapor is due to a reduction of free surface energy G^S of the condensed substance, equation (5.5) becomes:

$$-dG^S = RTd\ln P_S\,. \tag{5.7}$$

Upon integrating the given equation within the range from $G^S = 0$, P_S^*, when $A = A_0$, to G^S, P_S with A taking a certain value we have:

$$-G^S = RT\ln\frac{P_S}{P_S^*}. \tag{5.8}$$

In this equation G^S is equal to (see chapter 3):

$$G^S = -\sigma(A - A_0). \tag{5.9}$$

Here σ is the specific surface energy of the condensed substance.

With regard to the last two equations, we have:

$$\ln\frac{P_S}{P_S^*} = \frac{\sigma(A - A_0)}{RT}, \tag{5.10}$$

or if $A >> A_0$:

$$\ln\frac{P_S}{P_S^*} = \frac{\sigma A}{RT}. \tag{5.11}$$

If a nanodispersed substance consists of spherical particles of radius r in view of equation (3.3) we obtain:

$$\ln\frac{P_S}{P_S^*} = \frac{3\sigma V}{RTr}. \tag{5.12}$$

The equation above differs from the well-known Kelvin's equation by the factor of 3 instead of 2. This is due to the "three-dimensionality" of nanodispersity of spherical particles.

If a nanodispersed substance consists of strings of radius r or plates of thickness d (see section 3.1) we have the following equations:

$$\ln\frac{P_S}{P_S^*} = \frac{2\sigma V}{RTr}, \tag{5.13}$$

$$\ln\frac{P_S}{P_S^*} = \frac{2\sigma V}{RTd}. \tag{5.14}$$

The factor 2 in these equations implies that the nanodispersity of both strings and plates is "two-dimensional".

A comparison between equations (5.12) and (5.13) indicates that the saturated vapor pressure over spherical particles is higher that over the strings of the same radius r.

Table 5.1. Pressures of saturated vapor over nanosized drops of water, cadmium, and gold at various temperatures.

Temperature [K]	σ [J·m^{-2}]	P_S^* [bar]	P_S [bar] at r [m]		
			$50\cdot10^{-9}$	$5\cdot10^{-9}$	$0.5\cdot10^{-9}$
Water					
273.15	0.07564	$6.113\cdot10^{-3}$	$6.337\cdot10^{-3}$	$8.762\cdot10^{-3}$	0.2237
293.15	0.07275	0.02339	0.0242	0.0323	0.5925
313.15	0.06960	0.07381	0.0760	0.0988	1.3576
333.15	0.06624	0.19932	0.2046	0.2593	2.7631
353.15	0.06267	0.47373	0.4851	0.6007	5.0880
373.15	0.05891	1.01325	1.0352	1.2552	8.6258
Cadmium					
600	0.641	$1.79\cdot10^{-4}$	$1.99\cdot10^{-4}$	$5.30\cdot10^{-4}$	9.16
700	0.621	$3.43\cdot10^{-3}$	$3.76\cdot10^{-3}$	$8.52\cdot10^{-3}$	32.10
800	0.600	0.0308	0.0332	0.0674	80.08
900	0.581	0.167	0.1788	0.3320	160.82
Gold					
1338	1.170	$2.19\cdot10^{-8}$	$2.35\cdot10^{-8}$	$4.49\cdot10^{-8}$	$2.91\cdot10^{-5}$
1400	1.169	$8.69\cdot10^{-8}$	$9.31\cdot10^{-8}$	$1.73\cdot10^{-7}$	$8.42\cdot10^{-5}$
1500	1.168	$6.32\cdot10^{-7}$	$6.74\cdot10^{-7}$	$1.20\cdot10^{-6}$	$3.89\cdot10^{-4}$
1600	1.167	$3.58\cdot10^{-6}$	$3.80\cdot10^{-6}$	$6.54\cdot10^{-6}$	$1.48\cdot10^{-3}$
1700	1.166	$1.64\cdot10^{-5}$	$1.74\cdot10^{-5}$	$2.89\cdot10^{-5}$	$4.77\cdot10^{-3}$
1800	1.165	$6.37\cdot10^{-5}$	$6.72\cdot10^{-5}$	$1.09\cdot10^{-4}$	$1.36\cdot10^{-2}$

Let us apply equation (5.12) to determine the saturated vapor pressure over nanosized drops of some substances. In Table 5.1 the values of P_S^* and P_S are tabulated at radii of a drop $r = 50$, 5, and 0.5 nanometers for water, cadmium, and gold at various temperatures. The magnitudes of P_S^*, V, and σ necessary for calculation of P_S are taken from reference books [23,29,31] and Ref. [11].

An analysis of the data listed in Table 5.1 suggests the following. Firstly, saturated pressure vapor increases with a decrease in the diameter of drops of a liquid. Thus an appreciable increase in P_S as compared to P_S^* is observed only if the radius of a drop ranges from 5 up to 0.5 nm. Secondly, the higher is the temperature, the less significant is effect under consideration. For instance, at a temperature of 273.15K the ratio P_S/P_S^* for a drop of water, $r = 0.5$ nm in radius, is 36.6 times greater than that for bulk substance, while at 373.15K it increases only by a factor of 8.5. For cadmium the given ratio increases 51170 times at

600K and 980 times at 900K. For gold there are following values: 1330 times at 1338K and 213 times at 1800K.

Thirdly, a rise of saturated vapor pressure with increasing temperature is more appreciable for bulk substance than for nanosized drops. So, for water the saturated vapor pressure increases 156 times as the temperature increases from 273.15 to 373.15K, while for drops, r = 0.5 nm in radius, it rises 38 times. As temperature of cadmium increases from 600 up to 900K while for gold it changes from 1338 up to 1800K these values are equal to 933 and 17, 2908 and 467 respectively. This phenomenon results from the fact that the energy of evaporation of a liquid $\Delta_{vap}U$ decreases as both the surface area of a substance and its surface energy increase.

5.3. Dependence of phase transformation temperature of nanodispersed substance on the surface area

The phase transitions of the condensed substance include polymorphic transformation, melting (curve *adc* in Fig. 5.1), and the liquid-to-fluid transition (curve *bd* in Fig. 5.1).

Let us designate the free energies of two nanodispersed phases 1 and 2 of a given substance, which have the molar surface areas A_1 and A_2, as G_1^d and G_2^d. At phase transformation temperature of T_{tr} the phase 1 turns into the phase 2. This transition is characterized by the equality

$$G_1^d = G_2^d . \tag{5.15}$$

If the surface area of a phase 1 rises by dA_1 the phase transformation temperature will change by dT_{tr}. At the same time the surface area of a phase 2 will change by dA_2, while the increments in the free energies G_1^d and G_2^d will be $dG_1^d = dG_1^S$ and $dG_2^d = dG_2^S$. For phase transformation at temperature $T_{tr}+dT_{tr}$ the following equality is valid:

$$G_1^d + dG_1^S = G_2^d + dG_2^S . \tag{5.16}$$

Subtracting equation (5.15) from the present one we obtain the equality for phase transformation at temperature $T_{tr} + dT_{tr}$:

$$dG_1^S = dG_2^S . \tag{5.17}$$

With regard to the equations

$$dG_1^S = -S_1^S dT_{tr} - \sigma_1 dA_1 , \quad (5.18)$$

$$dG_2^S = -S_2^S dT_{tr} - \sigma_2 dA_2 \quad (5.19)$$

approximating $dA_l = dA_2 = dA$, equality (5.17) gives

$$\frac{dT_{tr}}{dA} = -\frac{\sigma_2 - \sigma_1}{S_2^S - S_1^S}. \quad (5.20)$$

In the equation above, S_2^S and S_1^S, σ_2 and σ_1 are the surface entropies and the specific surface energies of phases 1 and 2.

For the case of melting equation (5.20) can be rewritten:

$$\frac{dT_m}{dA} = -\frac{\sigma_L - \sigma_S}{S_L^S - S_S^S}. \quad (5.21)$$

From experience it is known that the melting temperature of any substance decreases as the surface area increases, i.e.:

$$\left(\frac{dT_m}{dA}\right)_P < 0. \quad (5.22)$$

It follows that for the substances characterized by $\sigma_S > \sigma_L$ (for instance, Au, Ag, Cu) inequality $S_S^S > S_L^S$ is observed, while for the substances with $\sigma_S < \sigma_L$ (for instance, Si, Bi) inequality $S_S^S < S_L^S$ is true.

From equation (3.3) for spherical particles we have:

$$dA = -\frac{3V}{r^2} dr.$$

Substituting the equation above into (5.21) we obtain:

$$\frac{dT_m}{dr} = \frac{3V(\sigma_L - \sigma_S)}{(S_L^S - S_S^S) r^2}. \quad (5.23)$$

For threadlike particles from equation (3.4) follows:

$$dA = -\frac{2V}{r^2} dr.$$

After substituting the equation above into (5.21) we have:

$$\frac{dT_m}{dr} = \frac{2V(\sigma_L - \sigma_S)}{(S_L^S - S_S^S) r^2}. \quad (5.24)$$

Let us introduce the designation

$$\alpha_m = \frac{\sigma_L - \sigma_S}{S_L^S - S_S^S} = \frac{\Delta_{fus}\sigma}{\Delta_{fus}S^S}, \tag{5.25}$$

as it contains but thermodynamic characteristics of melting, we shall name the quantity α_m the thermodynamic coefficient of melting. Then using equations (5.23) and (5.24) we shall obtain the following equations for spherical and threadlike particles:

$$\frac{dT_m}{dr} = \alpha_m \frac{3V}{r^2}, \tag{5.26}$$

$$\frac{dT_m}{dr} = \alpha_m \frac{2V}{r^2}. \tag{5.27}$$

The coefficient α_m for gold nanoparticles is calculated by the T_m,r curve (see fig.5.2) and equation (5.26) at various values r. Further, the magnitudes $\Delta_{fus}S^S$ are derived using the known values $\Delta_{fus}\sigma$ (see Tab. 4.2) and equation (5.25). The data are given in Table 5.2.

The data in Table 5.2 suggests that coefficient α_m increases as the radius of drops and melting temperature of gold decrease. It is basically due to a simultaneous decrease in the absolute magnitude of $\Delta_{fus}S^S$ (see equation (5.25)).

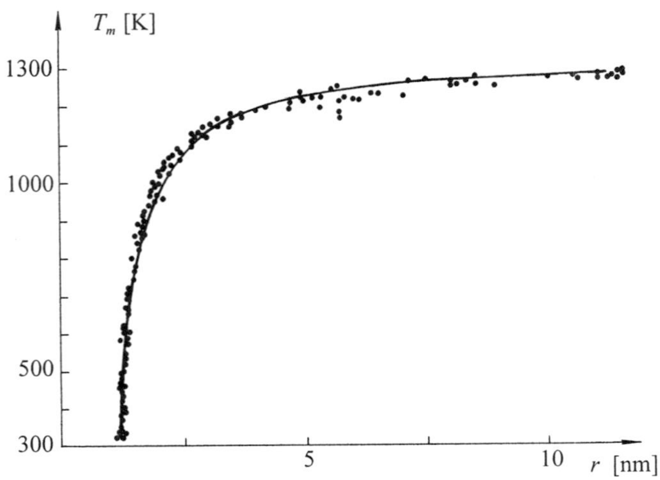

Fig. 5.2. Plot of melting temperature T_m as a function of radius of spherical nanoparticle of gold [10].

Table 5.2. Calculated values of coefficient α_m and the values of $\Delta_{fus}S^S$ for gold.

$r\cdot 10^{-9}$ [m]	T_m [K]	$(dT_m/dr)\cdot 10^{-9}$ [K·m^{-1}]	α_m [K·mol·m^{-2}]	$\Delta_{fus}\sigma$ [J·m^{-2}]	$\Delta_{fus}S^S$ [J·mol^{-1}·K^{-1}]
1.122	400	1821	0.0746	-0.033	-0.442
1.163	500	1363	0.0592	-0.034	-0.574
1.203	600	1091	0.0499	-0.035	-0.701
1.326	700	857	0.0470	-0.036	-0.766
1.488	800	633	0.0432	-0.036	-0.833
1.692	900	393	0.0342	-0.037	-1.082
2.081	1000	235	0.0307	-0.038	-1.238
2.961	1100	99	0.0256	-0.039	-1.523
4.160	1200	46	0.0234	-0.039	-1.667
6.700	1250	12.5	0.0164	-0.039	-2.378

5.4. Dependence of temperature and pressure at the first triple point on the surface area for nanodispersed substance. The first triple line.

As is well-known (see section 5.1), at the first triple point (point *a* in Fig. 5.1) the solid and the liquid states of a bulk substance are in equilibrium with the saturated vapor at pressure $P^*_S = P^*_{S-L}$ and temperature $T^*_m = T^*_{S-L}$ (see Fig.5.1). The surface area of a substance increasing, the temperature of the triple point T_{S-L} decreases, while the saturated vapor pressure P_{S-L} increases. The set of triple points T_{S-L} and P_{S-L} at various values A makes up the triple line P_{S-L}, T_{S-L}, A, along which saturated vapor is in equilibrium with both the solid and the liquid. If pressure P^*_{S-L} of saturated vapor is low at melting temperature T^*_{S-L}, it would be reasonable to assume that $T^*_m = T^*_{S-L}$ and $P^*_S = P^*_{S-L}$. Then the first triple line of any substance may be determined using the experimental dependence of melting temperature of the substance on radius of spherical nanoparticles.

The experimental dependence of melting temperature on radius of spherical gold nanoparticles is shown in Fig. 5.2. For gold $T^*_m = T^*_{S-L}$ = 1338K. At this temperature the saturated vapor pressure is $P^*_S = P^*_{S-L}$ = 2.19·10^{-8} bars [23], and A_0=0.00145 m^2·mol^{-1}.

The surface area of nanoparticles at known values of radius r can be found using the formula

$$A = \frac{3V}{r}. \tag{5.28}$$

The saturated vapor pressure for nanoparticles with known values A and melting temperatures $T_m = T_{S\text{-}L}$ may be obtained by the equation

$$\ln\frac{P_{S-L}}{P^*_{S-L}} = \frac{3\sigma V}{RT_m r}. \tag{5.29}$$

The values of σ and V at various values of T_m and r, which are necessary for calculation of $P_{S\text{-}L}$, are listed in Table 4.2. The obtained values $P_{S\text{-}L}$ at various A and $T_{S\text{-}L}$ are presented in Table 5.3 and in Figures 5.3 and 5.4.

The data below suggests that the saturated vapor pressure $P_{S\text{-}L}$ increases along the triple line $P_{S\text{-}L}, T_{S\text{-}L}, A$ with increasing surface area, while the melting temperature $T_{S\text{-}L}$ decreases (see Table 5.3, Fig. 5.3 and 5.4). Thus, when A_0 = 0.00145 $m^2 \cdot mol^{-1}$, $P^*_{S-L} = 2.19 \cdot 10^{-8}$ bar, and T^*_{S-L} = 1338K; and when A = 27352 $m^2 \cdot mol^{-1}$ $P_{S\text{-}L} = 1.98 \cdot 10^{-3}$ bar and $T_{S\text{-}L}$ = 400K. The projections of the first triple line $P_{S\text{-}L}, T_{S\text{-}L}, A$ for gold onto the planes $A, P_{S\text{-}L}$ and $A, T_{S\text{-}L}$ are shown in Fig. 5.3. Certain values of $P_{S\text{-}L}$ and $T_{S\text{-}L}$ (see vertical arrows in Fig. 5.3) correspond to each

Table 5.3. Coordinates of the first triple line $P_{S\text{-}L}, T_{S\text{-}L}, A$ for gold at various values r.

$r \cdot 10^{-9}$ [m]	σ [$J \cdot m^{-2}$]	A [$m^2 \cdot mol^{-1}$]	$T_{S\text{-}L} = T_m$ [K]	$P_{S\text{-}L}$ [bar]
1.122	1.264	27352	400	$1.98 \cdot 10^{-3}$
1.163	1.254	26551	500	$1.38 \cdot 10^{-4}$
1.203	1.244	25095	600	$2.41 \cdot 10^{-5}$
1.326	1.234	23704	700	$2.73 \cdot 10^{-6}$
1.488	1.224	21302	800	$1.45 \cdot 10^{-7}$
1.692	1.214	18892	900	$5.68 \cdot 10^{-7}$
2.081	1.204	15489	1000	$2.33 \cdot 10^{-7}$
2.961	1.194	10977	1100	$9.83 \cdot 10^{-8}$
4.160	1.184	7879	1200	$5.79 \cdot 10^{-8}$
6.700	1.179	4910	1250	$3.90 \cdot 10^{-8}$
$1.38 \cdot 10^{7}$*	1.170	0.00145*	1338	$2.19 \cdot 10^{-8}$

*Calculation has been performed under the assumption that the bulk gold is a sphere of volume V.

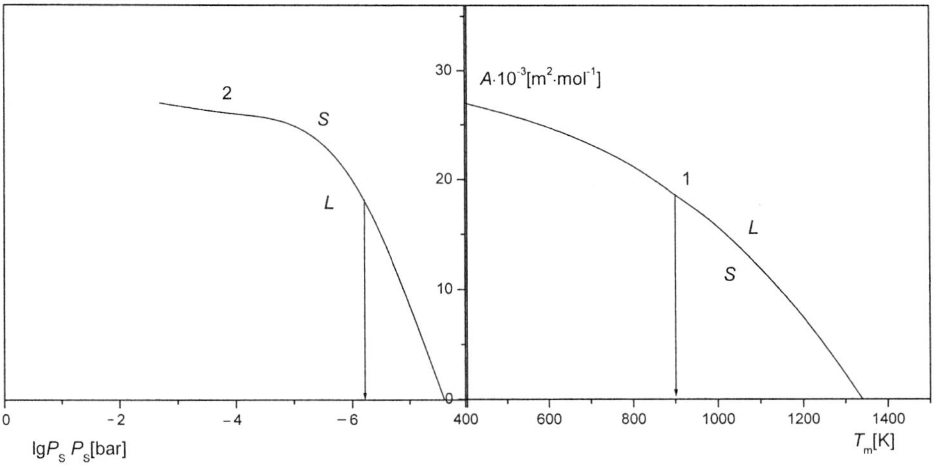

Fig. 5.3. Temperature of the first triple point $T_{S\text{-}L} = T_m$ (curve 1) and the pressure of saturated vapor $P_S = P_{S\text{-}L}$ at the first triple point (curve 2) versus the surface area A of spherical nanoparticles of gold.

value of A. The first triple line for gold related to the *PTA* coordinates is shown in Fig. 5.4 (curve aa_1). Clearly, the first triple line is a intersectional curve of planes aa_1cda and aa_1bka corresponding to equilibrium between the solid and the liquid gold with the saturated vapor. Fig. 5.4 shows that the region of a stable state of solid gold shifts towards lower temperatures and higher saturated vapor pressure with increasing surface (see plane aa_1cda in Fig. 5.4).

There are no experimental data which would enable us to construct the second triple line for some substance at the moment. Therefore this question will be not discussed in the current work.

5.5. On the distinction of pressure on each side of the curved surface of substance

Before approaching this issue let us dwell on the thermodynamic definition of pressure which results from the second law of thermodynamics at isothermal increase of pressure in differential form:

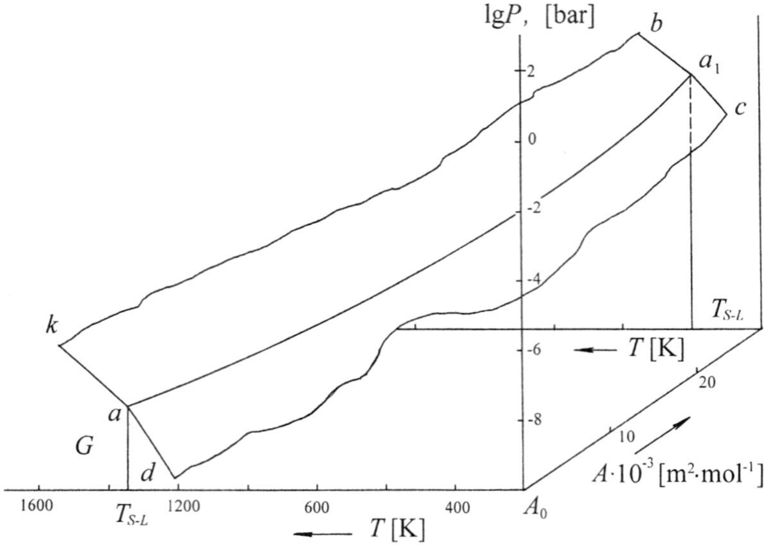

Fig. 5.4. The first triple line P_{S-L}, T_{S-L}, A for gold (curve aa_1). Curves *ad, ak*, and a_1c, a_1b are the temperature dependencies of saturated vapor pressure over the solid and the liquid gold at $A = A_0$ and $A = 27352$ m²·mol⁻¹. The aa_1cda and aa_1bka planes refer to the equilibrium pressures of saturated vapor over the solid and the liquid gold at various temperatures and magnitudes of A.

$$dU^P = PdE\ , \tag{5.30}$$

and integral form:

$$U^P - G^P = PE. \tag{5.31}$$

In equations (5.30) and (5.31) U^P and G^P are an increase in internal energy and decrease in free energy at buildup of external pressure from 0 to some value P, while E is the pressure coefficient of free energy defined by the equation:

$$E = -\left(\frac{\partial G^P}{\partial P}\right)_T . \tag{5.32}$$

From equation (5.31) follows:

$$P = \frac{U^P - G^P}{E}. \tag{5.33}$$

As is obvious, the ambient pressure on any substance is defined by the difference between energies U^P and G^P as well as by magnitude E according to the second law of thermodynamics. It is also independent of the surface form, for the equation of the second law of thermodynamics (5.31) does not depend on the structure of substance, its aggregative state, and the interface form between a substance and an environment. This means that at equilibrium the pressure on both sides of the phase boundary is the same and it does not depend on surface curvature.

In classical thermodynamics pressure P^α inside a spherical condensed substance is generally considered to exceed pressure P^β over the sphere by a magnitude ΔP, which depends on specific surface energy σ and radius r of a sphere:

$$\Delta P = P^\alpha - P^\beta = \frac{2\sigma}{r}. \tag{5.34}$$

The value of ΔP for a drop of water ($r = 0.5\cdot10^{-9}$ m) makes 2920 bar. Under the assumption that equation (5.34) is valid, such high pressure is built up without any work of compression at normal pressure over a drop. In order to increase pressure inside a drop up to 2920 bar it is necessary to increase external pressure up to this magnitude, for the pressure inside a substance with the curved or uncurved surface cannot be changed without a change in external pressure according to the second law of thermodynamics.

6

THERMODYNAMICS OF CAPILLARY FLOW OF LIQUID, PHYSICAL ADSORPTION, AND CAPILLARY CONDENSATION

If a solid is a capillary tube and comes in contact with a liquid, the latter can spontaneously rise in this capillary tube to a certain height determined by the interaction of the liquid with the surface of the solid. If the surface of solid comes into contact with the vapor of a liquid, there can appear a thin layer of the liquid (adsorbate) on it. This phenomenon is referred to as physical adsorption. In the present chapter the thermodynamics of capillary flow of a liquid, physical adsorption, and capillary condensation is discussed.

6.1. Wetting of solid surface by a liquid. Contact angle

The contact of a drop of a liquid with a solid leads to a change in the drop form, owing to the interaction between the solid surface and the liquid, for the liquid forms a certain surface area at the boundary with the solid. The process of spontaneous increase in the surface area of a liquid at the boundary with the solid is referred to as the wetting of solid surface by a liquid or, simply, wetting. In equilibrium between a drop of a liquid and a solid surface a certain angle θ is established, which is called the contact angle (wetting angle). Depending on the magnitude of an contact angle the following cases of wetting are distinguished (see Fig.6.1):

- at $\theta < 90°$ ($\cos\theta > 0$) the liquid does wet the solid surface (Fig. 6.1, a),
- at $\theta > 90°$ ($\cos\theta < 0$) the liquid does not wet the solid surface well (negative wetting) (Fig. 6.1, c),
- at $\theta = 0°$ the liquid wets the solid surface completely (absolute wetting) (Fig. 6.1, b),
- at $\theta = 180°$ the liquid absolutely does not wet the solid surface (absolute non-wetting) (Fig. 6.1, d).

Thus, the smaller is the contact angle, the better the liquid wets a solid surface. The

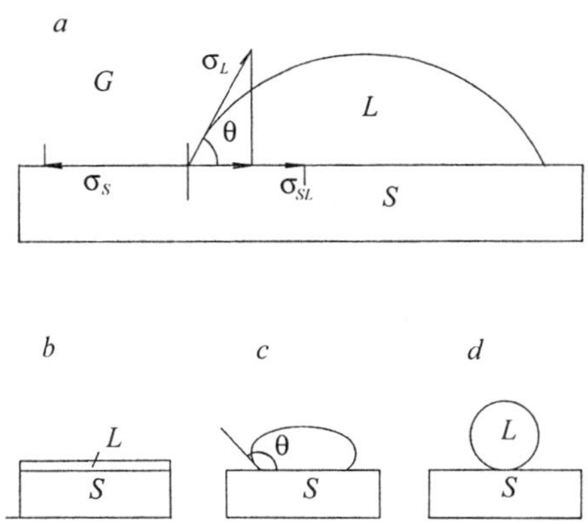

Fig. 6.1. Drop of liquid L on the surface of solid S in the atmosphere of gas G. a) θ – contact angle, σ_S and σ_S – the specific surface energy of a solid and a liquid at the border with gas, σ_{SL} – the specific interfacial energy. Three situations of wetting the surface of a solid by various liquids: b) formation of a film at absolute wettability, c) wetting at $\theta > 90°$, d) formation of a spherical drop at absolute non-wettability.

values of θ for some liquids at the boundary with various solids are listed in Table 6.1.

The data in Table 6.1 suggest that water does not wet wool well, wets the nickel surface well, and wets the glass surface absolutely. Mercury wets both glass and steel insufficiently.

6.2. Specific interfacial energy and specific energy of wetting

When a liquid wets a solid, their specific surface energies σ_L and σ_S change and, as a result, there is a specific interfacial energy σ_{SL} at the phase boundary. The quantity σ_{SL} can be seen as the force which acts at the liquid/solid interface and tends to reduce the interfacial area. Thinking of the quantity σ_S as the force which operates on the solid surface and tends to increase the interfacial area, and with the horizontal projection of force σ_L taking into account, the balance of forces in a horizontal plane (see Fig.6.1, a) gives the equation

$$\sigma_S = \sigma_{SL} + \sigma_L \cos\theta, \tag{6.1}$$

which has been labeled Young equation and also occurs in the following form:

$$\cos\theta = \frac{\sigma_S - \sigma_{SL}}{\sigma_L}, \tag{6.2}$$

Table 6.1. Contact angle θ for some liquid–solid systems at room temperature.

Liquid	Solid	Contact angle θ [°]
Water	Wool	160
Water	Teflon	120
Water	Nickel	27
Water	Glass	0
Water	Steel	154
Water	Glass	140
Benzene	Glass	6
Oleic acid	Glass	30

$$\sigma_L \cos\theta = \sigma_L - \sigma_{SL}. \qquad (6.3)$$

The quantity σ_{SL} is unknown in these equations. Therefore we shall introduce a new quantity, namely, the specific energy of wetting σ^w, which is defined by the equation:

$$\sigma^w = \sigma_L - \sigma_{SL}. \qquad (6.4)$$

From equations (6.1) and (6.4) follows:

$$\sigma^w = \sigma_L \cos\theta \qquad (6.5)$$

or

$$\cos\theta = \frac{\sigma^w}{\sigma_L}. \qquad (6.6)$$

From equation (6.5) follows that specific energy of wetting can be both a positive and a negative magnitude. When $\theta < 90^0$ $\cos\theta > 0$ and hence $\sigma^w > 0$; when $\theta = 90^0$ $\cos\theta = 0$ and $\sigma^w = 0$, and at $\theta > 90^0$ $\cos\theta < 0$ and $\sigma^w < 0$. The preceding is illustrated by Fig. 6.2.

6.3. Rising and lowering of liquid in a capillary and the capillary energy of wetting

If wetting the wall surface of a capillary, a liquid is known to spontaneously flow into a capillary and rise to the certain height h (see Fig.6.3, a). If the liquid does not wet the wall surface of a capillary well, it goes down to the certain height h in the capillary (Fig. 6.3, b). The height reached by the liquid as it rise or goes down in a capillary depends on the specific surface energy σ_L of a liquid, contact angle θ, and radius of a capillary r.

Parallelograms of the forces equal to magnitudes σ_S, σ_L, σ_{SL}, and $\sigma_L\cos\theta$ are presented in Fig. 6.4 (a and b). The balance of forces along a capillary wall, characterizing the rise of the liquid, can be described by the equations:

$$\sigma_S = \sigma_{SL} + \sigma_L \cos\theta, \qquad (6.7)$$

$$\sigma_S = \sigma_{SL} + \sigma^w. \qquad (6.8)$$

If the specific energy of wetting $\sigma^w > 0$, the liquid rises in a capillary above a level of a liquid in a vessel (see Fig. 6.3, a). If $\sigma^w < 0$ the liquid falls below a level

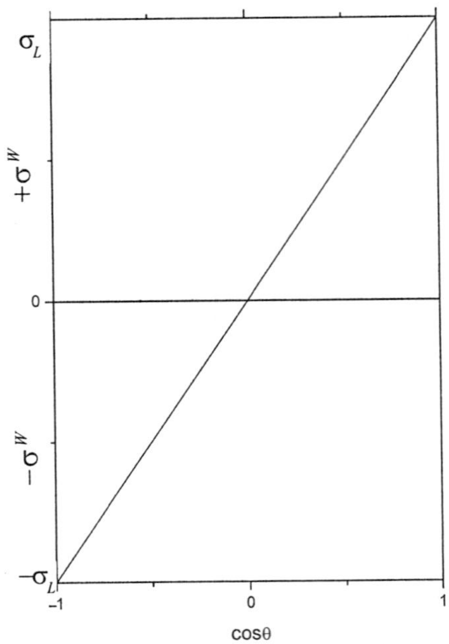

Fig. 6.2. The specific energy of wetting σ^w versus cosine of contact angle θ.

of a liquid in a vessel (see Fig.6.3, b). Energy of liquid wetting u^w of a capillary equals to the product of specific energy of wetting σ^w and the surface area a^w of the part of a capillary in height h equal to height of rise or lowering of a liquid in a capillary:

$$u^w = \sigma^w a^w . \tag{6.9}$$

Taking into account $a^w = 2\pi rh$ (r is the radius of a capillary, h is the height of liquid raising or lowering in a capillary) and equation (6.5), we find:

$$u^w = 2\pi rh\sigma_L \cos\theta . \tag{6.10}$$

The quantity u^w refers to one capillary. Consequently, it may be named the capillary energy of wetting.

6.4. Dependence of height of liquid rising and lowering in capillary on its radius

When wetting the walls of a capillary, a liquid rises to a certain height h. The formed liquid column of a volume $v = \pi r^2 h$ produces the hydrostatic pressure $p = gh(\rho - \rho_g)$. Equilibrium is established when the work of liquid rising in a capillary:

$$w = vp = \pi r^2 h^2 q(\rho - \rho_g) \tag{6.11}$$

becomes equal to the capillary energy of wetting u^w.

Equating (6.10) and (6.11), we have the required equation

$$h = \frac{2\sigma_L \cos\theta}{g(\rho - \rho_g)r} . \tag{6.12}$$

In equations (6.10) and (6.11) $g = 9.80665$ m·s^{-2} is the acceleration of gravity, ρ and ρ_g are the liquid and gas densities.

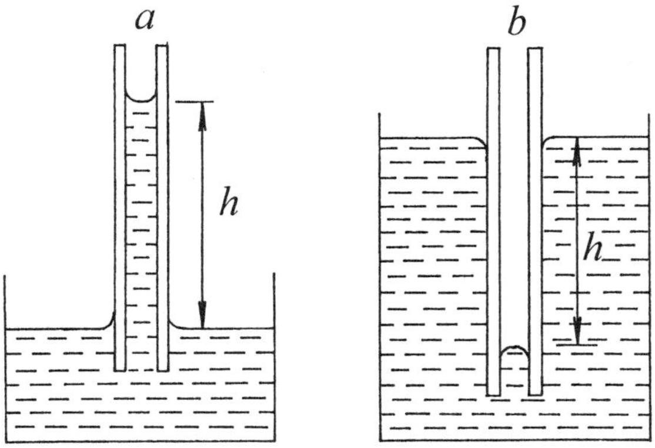

Fig. 6.3. Rising (a) and lowering (b) of liquid in a capillary: h – height of a rising or lowering of a liquid relative to the level of a liquid in a vessel.

From equation (6.12) results the conclusion that the quantity h depends inversely only on the radius of a capillary for the chosen system liquid / solid. The magnitude h can be calculated by equation (6.12) if magnitudes σ_L, θ, ρ, ρ_g, and r are known. It will be positive if $\theta < 90^0$ ($\cos\theta > 0$) and negative if $\theta > 90^0$ ($\cos\theta < 0$).

In equation (6.12) the weight of a meniscus part of a liquid column is not taken into account. To take it into account the form of a meniscus has to be known. As a first approximation, the meniscus is assumed to be a concave or convex hemisphere with a radius of r. In this case, the estimated mistake does not exceed several tenths of one percent of h. Consequently, we shall use equation (6.12) henceforth.

Let us rewrite equation (6.12) in the following way:

$$hr = \frac{2\sigma_L \cos\theta}{g(\rho - \rho_g)}. \qquad (6.13)$$

In this equation the quantity

$$a^2 = \frac{2\sigma_L}{g(\rho - \rho_g)} \qquad (6.14)$$

depends only on the properties of a liquid. Consequently, the coefficient a will be denoted as the capillary constant of a liquid. It can be calculated by equation (6.14). Then equation (6.13) becomes

$$hr` = a^2\cos\theta. \qquad (6.15)$$

Apparently, the product *hr* for the liquid depends only on a contact angle. Taking into account (6.15) and using (6.10), we have the equation for determining the capillary energy of wetting:

$$u^w = 2\pi a^2 \sigma_L (\cos\theta)^2. \tag{6.16}$$

Thus, the capillary energy of wetting u^w can be determined if the capillary constant a of a liquid, its specific surface energy σ_L, and the contact angle θ are known.

6.5. Capillary free energy of wetting

As is well known (see sections 3.3 and 3.5), the surface energy U^S and the surface free energy G^S are equal in absolute magnitude but opposite in sign:

$$U^S = -G^S. \tag{6.17}$$

Therefore for a capillary the following equality is true:

$$u^w = -g^w, \tag{6.18}$$

where g^w is the capillary free energy of wetting. With regard to equality (6.18) the magnitude g^w can be calculated with the help of equation (6.16).

6.6. Capillary entropy of wetting

The capillary entropy of wetting s^w is the temperature coefficient of the capillary free energy of wetting:

$$s^w = -\left(\frac{\partial g^w}{\partial T}\right)_{P,a^w}. \tag{6.19}$$

The magnitude g^w can be calculated by the equation

$$g^w = -\sigma^w a^w. \tag{6.20}$$

In the equation the quantity σ^w depends on temperature, since both the specific surface energy of a liquid σ_L and $\cos\theta$ in equation (6.5) depend on temperature. Temperature dependence of the area of wetting a^w is determined by the temperature dependence of the product rh:

$$rh = \frac{a^w}{2\pi}. \qquad (6.21)$$

The magnitude rh can be determined by equation (6.13).

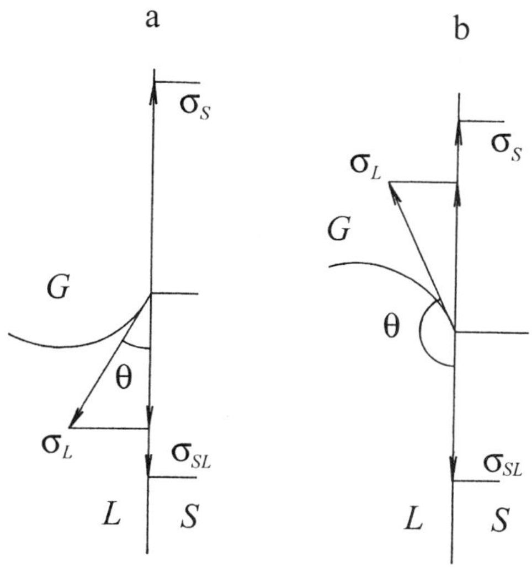

Fig. 6.4. Effective forces at the three-phase border liquid (L) – solid (S) – gas (G) at rising (a) and lowering (b) of liquid in a capillary: σ_S and σ_L are the specific surface energies of solid and liquid at the border with gas, σ_{SL} is the specific interfacial energy, θ is a contact angle.

6.7. Molar area of wetting

The volume of a liquid v in a capillary and the capillary area of wetting a^w at $h > 0$ are equal to

$$v = \pi r^2 h, \qquad (6.22)$$

$$a^w = 2\pi r h. \qquad (6.23)$$

Hence it follows that the number of capillaries z in which there is one mole of a liquid with a volume of V equals

$$z = \frac{V}{\pi r^2 h}. \qquad (6.24)$$

Considering that $A^w = za^w$, the molar area of wetting A^w may be expressed by the equation

$$A^w = \frac{2V}{r}, \qquad (6.25)$$

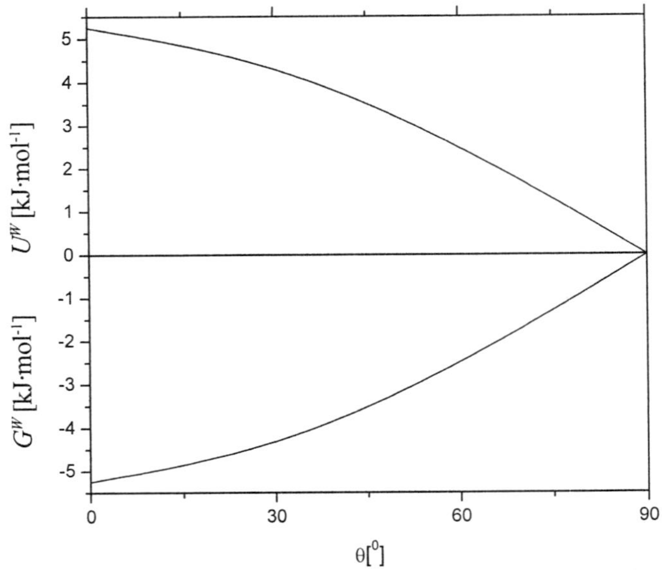

Fig. 6.5. Plot of energy U^w and free energy G^w of wetting for water at 20 °C as a function of contact angle θ at radius of a capillary $r = 0.5$ nm.

from which follows that the magnitude A^w is inversely proportional to the radius of a capillary r.

6.8. Energy and free energy of wetting

As the specific energy of wetting $\sigma^w = \sigma_L \cos\theta$ the energy of wetting U^w is equal

$$U^w = \sigma^w A^w = \frac{2V\sigma_L \cos\theta}{r}. \tag{6.26}$$

With regard to $U^w = -G^w$ for the free energy of wetting G^w we obtain the equation

$$G^w = -\frac{2V\sigma_L \cos\theta}{r}. \tag{6.27}$$

Fig. 6.5 shows the dependence of U^w and G^w on θ for water in the capillary of radius $r = 0.5$ nm when $\theta = 0 - 90^0$.

6.9. Entropy of wetting

The molar entropy of wetting is equal to

$$S^w = s^w z \tag{6.28}$$

or in view of equation (6.24), we find:

$$S^w = s^w \frac{V}{\pi r^2 h}. \tag{6.29}$$

As pointed out, for determining the capillary entropy of wetting s^w the temperature dependence of the capillary free energy of wetting g^w has to be known.

6.10. Physical adsorption and capillary condensation

6.10.1. Experimental data

Consider a sample 1 of a solid in an evacuated vessel (see Fig. 6.6) in which the pressure P_S of vapor G of liquid L being built up is less than the saturated vapor pressure P_S^* at the chosen temperature ($P_S < P_S^*$). Under the condition that a liquid wets the surface of solid the following phenomena are observed:

- if the solid sample does not possess porosity, the nanophase of the condensed liquid (adsorbate) develops on its surface. The monolayer and then multilayer of nanophase of adsorbate are gradually formed on the surface of solid with increase of pressure P_S of the vapor. The phenomenon described is referred to as physical adsorption;
- if the sample is a nanoporous material, nanopores are gradually filled up by the liquid (adsorbate) with increased pressure of vapor P_S. In this case the liquid nanophase also appears. This phenomenon is referred to as capillary condensation.

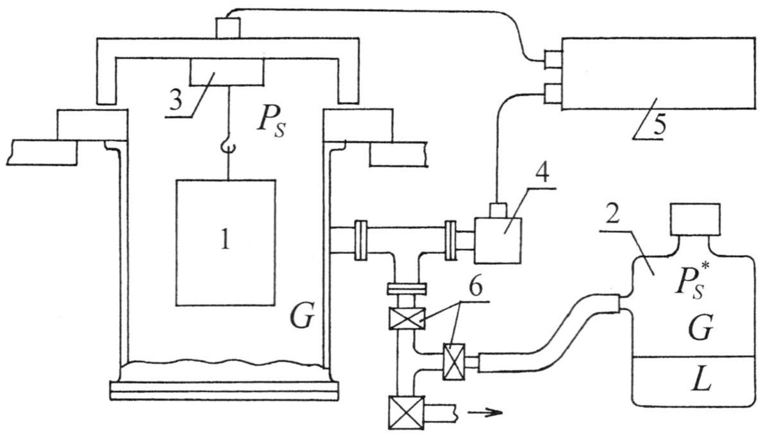

Fig. 6.6. Diagram of unit for determination of isotherm of adsorption: L – liquid, G – vapor, 1 – sample, 2 – vessel with liquid and saturated vapor, 3 – weighing cell, 4 – pressure sensor, 5 – recorder, 6 – valves.

In both cases equilibrium between the vapor and the liquid nanophase is established. An amount of nanophase *m(ads)* and its size (thickness of the adsorbate layer, size of the pore filled by a liquid) increase (Fig. 6.7) as the P_S/P_S^* ratio increases. Curves 1 and 2 in Fig. 6.7 are the typical isotherms of gas adsorption on the surface of non-porous and nanoporous solids.

Let us turn to experiment. Curve 1 in Fig. 6.8 shows equilibrium between relative pressure P_S/P_S^* of nitrogen at –196 °C and the number of adsorbate monolayers on the surface of non-porous silica. As appears, the relative pressure of nitrogen increases with an increase in thickness of adsorbate nanophase.

For the sake of comparison, the dependence of the relative pressure of nitrogen on the radius of a liquid nitrogen drop at –196 °C is shown in Fig. 6.8 (curve 2). The relative pressure P_S/P_S^* has been calculated according to the equation

$$\ln\frac{P_S}{P_S^*} = \frac{3\sigma_L V}{RTr}, \quad (6.30)$$

where $\sigma_L = 0.00892$ J·m^{-2}, $V = 34.576\cdot10^{-6}$ m^3·mol^{-1}.

A comparison between curves 1 and 2 in Fig. 6.8 makes it clear that in the second case the relative pressure P_S/P_S^* decreases with an increase in nanophase size, while the inverse effect is observed at the adsorption of nitrogen on the surface of a solid: the relative pressure increases with an increase in thickness of adsorbate nanophase. The problem is discussed in detail below.

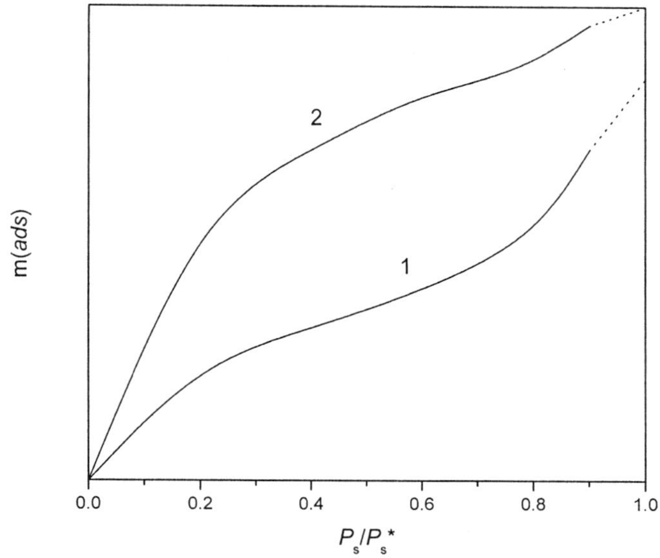

Fig. 6.7. Typical isotherms of adsorption on the surface of non-porous (1) and nanoporous (2) solids.

6.10.2. Internal and free energies of adsorbate and liquid

As is well known, the following equations are valid for a nanodispersed liquid:

$$U^d = U + U^S, \tag{6.31}$$

$$G^d = G + G^S, \tag{6.32}$$

where U^d and G^d, U and G are the internal and free energies of nanodispersed and bulk liquids, U^S and G^S are the surface and free surface energies of nanodispersed liquid. They are defined by the equations

$$U^S = \sigma_L A, \tag{6.33}$$

$$G^S = -\sigma_L A, \tag{6.34}$$

where σ_L and A are the specific surface energy of a liquid and the molar surface area.

In the case under consideration the relative pressure P_s/P_S^* and the surface free energy G^S of a liquid are related by the equation

$$\ln\frac{P_S}{P_S^*} = -\frac{G^S}{RT}. \tag{6.35}$$

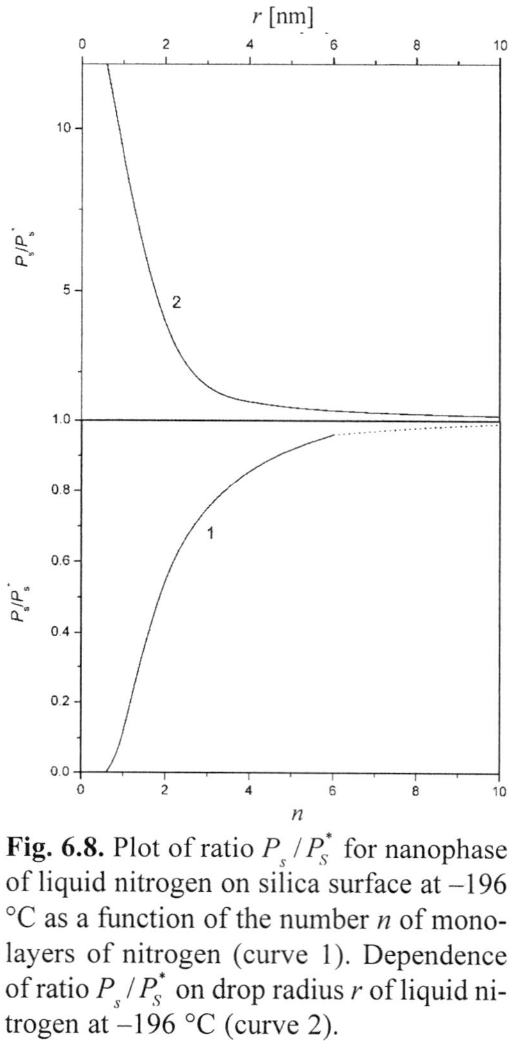

Fig. 6.8. Plot of ratio P_s / P_S^* for nanophase of liquid nitrogen on silica surface at –196 °C as a function of the number n of monolayers of nitrogen (curve 1). Dependence of ratio P_s / P_S^* on drop radius r of liquid nitrogen at –196 °C (curve 2).

$G^S < 0$ since $P_s / P_S^* > 1$, i.e. the pressure of saturated vapor of nanophase P_s exceeds the pressure of saturated vapor P_S^* of liquid macrophase if $G^S < 0$. In this case $U^S > 0$.

For adsorbate nanophase $P_s / P_S^* < 1$ at adsorption and capillary condensation. Consequently, the surface free energy $G^S(ads)$ of adsorbate is positive ($G^S(ads) > 0$), and its surface energy $U^S(ads)$ is negative ($U^S(ads) < 0$).

From the equations

$$U^d(ads) = U + U^S(ads), \quad (6.36)$$

$$G^d(ads) = G + G^S(ads) \quad (6.37)$$

follows:

$$U^S(ads) = U^d(ads) - U, \quad (6.38)$$

$$G^S(ads) = G^S(ads) - G. \quad (6.39)$$

As $U^S(ads) < 0$, from equation (6.38) follows that the internal energy of adsorbate is less than that of a liquid macrophase. In other words, when a liquid goes from a macrophase to nanophase on the surfaces of a solid its internal energy decreases with free energy increasing. Fig.6.9 schematically illustrates it. The similar situation is observed at capillary condensation.

6.10.3. Saturated vapor pressure of adsorbate

The relative pressure P_s / P_S^* is related to the free surface energy of adsorbate by the equation

$$\ln \frac{P_S}{P_S^*} = -\frac{G^S(ads)}{RT}. \quad (6.40)$$

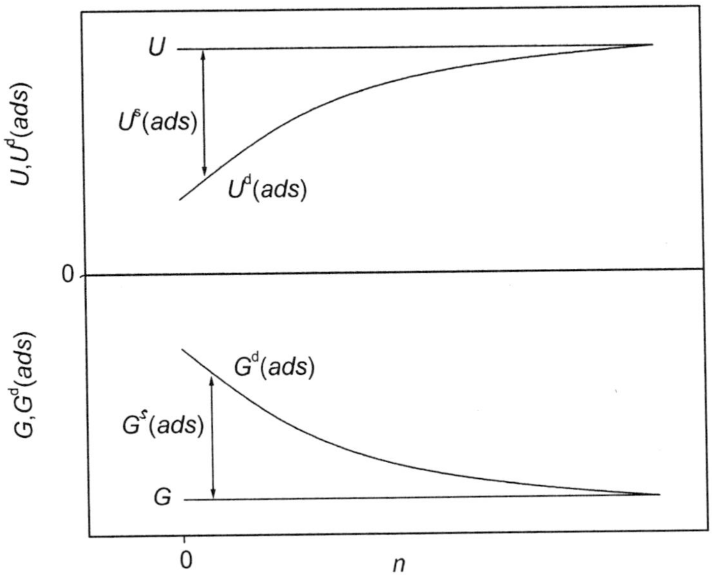

Fig. 6.9. Change in the internal energy $U^d(ads)$ and free energy $G^d(ads)$ of adsorbate nanophase with an increase in number n of monolayers of adsorbate at physical adsorption: U, G are the internal and free energies of liquid macrophase; $U^S(ads)$, $G^S(ads)$ are the surface energy and surface free energy of adsorbate (diagram).

Since $G^S(ads) > 0$, in accordance with experiment $P_s/P_S^* < 1$ (see curve 1 in Fig. 6.8).

Equation (6.40) is true for both adsorption and capillary condensation. Therefore using this equation makes possible to find the magnitude $G^S(ads)$, provided that the dependence of ratio P_s/P_S^* on an amount of adsorbate (for example, on the number of moles of adsorbate on a surface or in pores of a sample) is known.

6.10.4. Thermodynamic characteristics of adsorbate at physical adsorption

As stated above, when a liquid macrophase B passes into a nanophase B*(ads)* or an adsorbate:

$$B = B(ads) \tag{6.41}$$

the free energy of a liquid raises by the amount $G^S(ads)$, and its internal energy decreases by $U^S(ads)$.

The magnitude $G^S(ads)$ can be determined using experimental dependence of adsorbate mass on relative pressure P_s / P_S^*. After having determined magnitudes $G^S(ads)$ at various values $m(ads)$ we construct the $G^S(ads),m(ads)$ curve. The dependence of number n of adsorbate monolayers on relative pressure is frequently determined at adsorption. In this case we shall construct curve $G^S(ads),n$.

Since

$$U^S(ads) = -G^S(ads), \tag{6.42}$$

the $U^S(ads),m(ads)$ (or n) curve can be constructed with the help of the $G^S(ads),m(ads)$ (or n) curve.

With the help of curve 1 in Fig. 6.8 and the procedure stated above the magnitudes $G^S(ads)$ for nitrogen are determined at its adsorption on silica surface at

Table 6.2. Free surface energy $G^S(ads)$, surface energy $U^S(ads)$, area A, and specific surface energy of adsorbate $\sigma(ads)$ at adsorption of nitrogen on the silica surface at –196 °C.

P_s / P_S^*	n	$G^S(ads)$ [J·mol^{-1}]	$U^S(ads)$ [J·mol^{-1}]	A [m^2·mol^{-1}]	$\sigma(ads)$ [J·m^{-2}]
0.005	0.562	3392	-3392	97763	-0.0347
0.01	0.687	2948	-2948	97763	-0.0301
0.025	0.812	2362	-2362	97763	-0.0241
0.05	0.938	1914	-1914	97763	-0.0196
0.075	1.00	1658	-1658	97763	-0.0169
0.1	1.11	1472	-1472	88075	-0.0167
0.2	1.25	1031	-1031	78210	-0.0132
0.3	1.39	771	-771	70333	-0.0110
0.4	1.53	586	-586	63897	-0.0092
0.5	1.7	444	-444	57507	-0.0077
0.6	1.94	327	-327	50393	-0.0065
0.7	2.22	229	-229	44037	-0.0052
0.8	2.81	143	-143	34791	-0.0041
0.9	3.87	67	-67	25262	-0.0026
0.95	5.5	33	-33	17775	-0.0018
0.975	6.5	16	-16	15040	-0.0011

−196 °C. These data at various values of both relative pressure and number of adsorbate monolayers n are listed in Table 6.2.

To determine the specific surface energy of adsorbate $\sigma(ads)$ we shall use the equation

$$G^S(ads) = -\sigma(ads)A. \qquad (6.43)$$

where A is the molar area of adsorbate at the boundary with solid. It depends on the number of adsorbate monolayers at $n \geq 1$:

$$A = \frac{A_m}{n}. \qquad (6.44)$$

In this equation A_m is the monomolecular area of adsorbate (the surface area of one mole of adsorbate on the surface of a solid). The magnitude A_m can be calculated by the equation [12]:

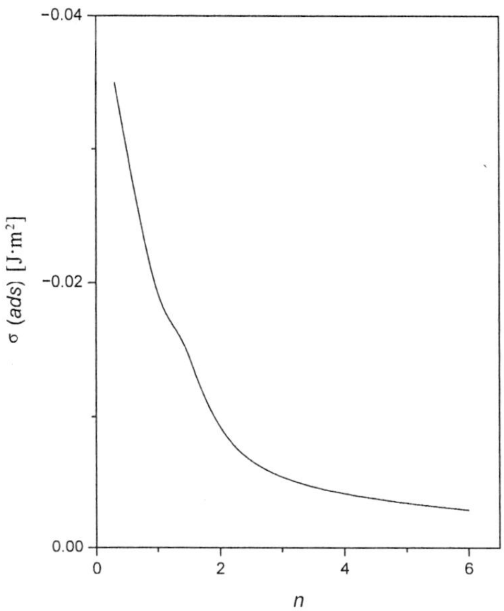

Fig. 6.10. Plot of specific surface energy $\sigma(ads)$ of adsorbate (nitrogen) as a function of the number n of monolayers of nitrogen on silica surface at −196 °C.

$$A_m = f\left(\frac{V}{N}\right)^{2/3} N, \qquad (6.45)$$

where V is the molar volume of a liquid, $N = 6.022 \cdot 10^{23}$ is the Avogadro's constant, and f is the packing factor which is usually assumed to be equal to 1.09 [12].

For nitrogen at −196 °C $V = 34.576 \cdot 10^{-6}$ m^3·mol^{-1}. Consequently, in this case we shall obtain A_m = 97763 m^2·mol^{-1} with the help of equation (6.45). If the number of adsorbate monolayers is less than or equal to unity ($n \leq 1$), the molar area of adsorbate $A = A_m$. The magnitudes A and $\sigma(ads)$ found with the help of the equations (6.44) and (6.43) are shown in Table 6.2. The $\sigma(ads),n$ dependence is displayed in Fig. 6.10.

It is evident from Fig.6.10 that the absolute value of the specific surface energy of adsorbate $\sigma(ads)$ decreases sharply after the first monolayer is filled up. This suggests that the interaction of adsorbate molecules among themselves has a dramatic effect on the magnitude $\sigma(ads)$. After the first adsorbate monolayer is filled the reduction of $\sigma(ads)$ becomes weaker as monolayers grow in number.

After the second monolayer is filled the further reduction goes rather slowly (see Table 6.2 and Fig. 6.10).

6.10.5. Thermodynamic characteristics of adsorbate at gas adsorption by nanoporous solid

In this case the v_A, P_s/P_S^* curve is found practically, where v_A is the volume of adsorbate per 1g of solid. Equation (6.40) makes it possible to calculate the surface free energy $G^S(ads)$ of the adsorbate lying in nanopores as well as on their walls.

Fig. 6.11 shows the isotherm of both adsorption and capillary condensation of nitrogen on saran coal at –195.6 °C. The calculated values $G^S(ads)$ at various values of volume v_A of nitrogen in sample pores are shown in Table 6.3. As is obvious, the free surface energy $G^S(ads)$ decreases as the volume of the absorbed nitrogen increases. This fact is due to an increase in the amount of adsor-

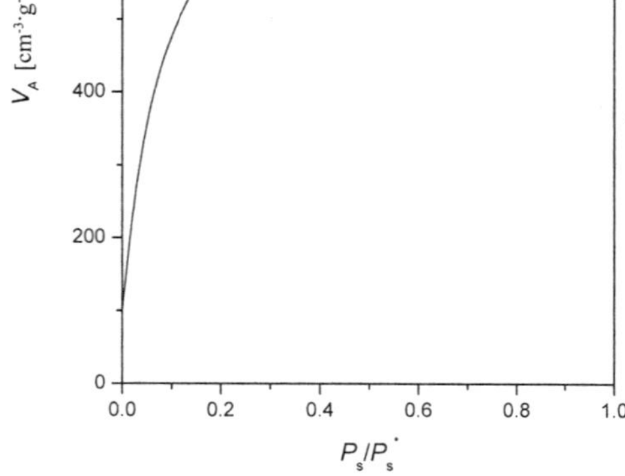

Fig. 6.11. Isotherm of the adsorption of nitrogen on a sample of saran coal at –195.6 °C.

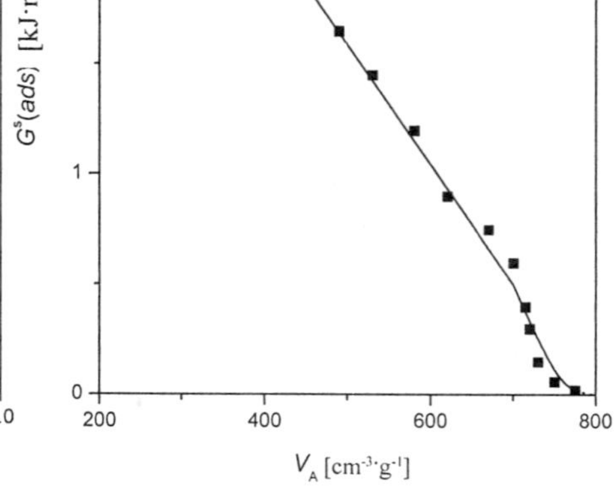

Fig. 6.12. Plot of $G^S(ads)$ versus the volume v_A of nitrogen adsorbed by nanoporous saran coal at –195.6 °C (see fig. 6.11).

Table 6.3. Free surface energies $G^S(ads)$ and surface energies $U^S(ads)$ at adsorption of nitrogen by nanodispersed coal at –195.6 °C.

P_s / P_S^*	v_A [cm³·g⁻¹]	$G^S(ads)$[J·mol⁻¹]	$U^S(ads)$[J·mol⁻¹]
0.010	225	2935	-2935
0.025	360	2351	-2351
0.050	440	1909	-1909
0.075	495	1650	-1650
0.10	530	1467	-1467
0.15	580	1209	-1209
0.20	620	1025	-1025
0.25	650	883	-883
0.30	675	767	-767
0.40	705	584	-584
0.50	715	442	-442
0.60	720	325	-325
0.70	725	227	-227
080	730	142	-142
090	745	67	-67
0.95	775	33	-33

bate contacting with the walls of pores. The values of $U^S(ads)$ based on the equality $U^S(ads) = -G^S(ads)$ are also listed in Table 6.3.

Fig. 6.12 shows the $G^S(ads)$ dependence on v_A. As is evident, the $G^S(ads),v_A$ dependence is linear over a wide interval of values v_A, i.e. it is described by the equation

$$G^S(ads) = av_A, \tag{6.46}$$

where a is a constant of proportionality which is equal in this case to –4.88 J·g·mol⁻¹·cm⁻³.

Let us turn to the well-known Kelvin equation

$$\ln\frac{P_S}{P_S^*} = -\frac{2\sigma_L V\cos\theta}{RTr}, \tag{6.47}$$

where σ_L is the specific surface energy of liquid adsorbate, V is the molar volume, θ is the contact angle, and r is the radius of a pore.

When passing from the equation

$$\ln\frac{P_S}{P_S^*} = -\frac{G^S(ads)}{RT}, \tag{6.48}$$

which defines the quantity of relative pressure at various values of $G^S(ads)$, to equation (6.47) the equality

$$G^S(ads) = \frac{2\sigma_L V\cos\theta}{r} \tag{6.49}$$

has to be assumed valid.

There is no experimental data on the equality validity.

7

THERMODYNAMIC PROPERTIES OF BINARY NANODISPERSED SOLUTIONS

Thermodynamic properties of binary solutions A–B were discussed in detail in Ref. [8]. In this chapter the passing from the thermodynamic properties of bulk to the properties of nanodispersed solution is considered.

7.1. Thermodynamic properties of binary solution

7.1.1. Thermodynamic properties of ideal solution

The thermodynamic properties of a bulk solution – the internal energy U_{sln}, the free energy G_{sln}, and the entropy S_{sln} are generally made up of the thermodynamic properties of an ideal solution U^{id}_{sln}, G^{id}_{sln}, and S^{id}_{sln} and those of mixing $\Delta_{mix}U$, $\Delta_{mix}G$, and $\Delta_{mix}S$:

$$U_{sln} = U^{id}_{sln} + \Delta_{mix}U, \tag{7.1}$$

$$G_{sln} = G^{id}_{sln} + \Delta_{mix}G, \tag{7.2}$$

$$S_{sln} = S^{id}_{sln} + \Delta_{mix}S. \tag{7.3}$$

For an ideal solution we have: $\Delta_{mix}U = \Delta_{mix}G = \Delta_{mix}S = 0$.

The magnitudes of U^{id}_{sln}, G^{id}_{sln}, and S^{id}_{sln} are calculated using the values U_A and U_B, G_A and G_B, S_A and S_B which are known for pure substances [20 – 25] according to the equations:

$$U^{id}_{sln} = U_A + x_B(U_B - U_A), \tag{7.4}$$

$$G^{id}_{sln} = G_A + x_B(G_B - G_A), \tag{7.5}$$

$$S^{id}_{sln} = S_A + x_B(S_B - S_A), \tag{7.6}$$

where x_B is the mole fraction of substance B in a solution A–B.

As a case in point we consider the Ag–Au binary liquid solution at 1379K. Let us calculate its thermodynamic properties as those of an ideal solution. Those of silver (A) and gold (B) at $T = 1379$K are listed below.

Substance	U [kJ·mol^{-1}]	G [kJ·mol^{-1}]	S [J·mol^{-1}·K^{-1}]
Ag	48.276	-82.079	94.529
Au	48.649	-84.837	96.799

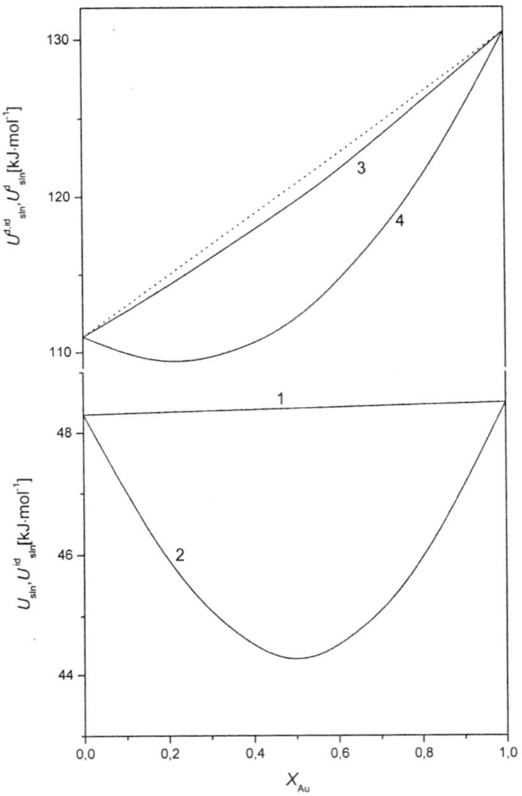

Fig. 7.1. Plot of the internal energy of bulk (curves 1 and 2) and nanodispersed (curves 3 and 4) solutions of Ag–Au as a function of x_{Au} at 1379K. Curves: 1 – U^{id}_{sln}, x_{Au}, 2 – U_{sln}, x_{Au}, 3 – $U^{d,id}_{sln}$, x_{Au}, 4 – U^{d}_{sln}, x_{Au}.

Using the data above as well as equations (7.4) and (7.6) we obtain the following equations allowing to calculate the thermodynamic properties of the ideal binary solution Ag–Au:

$$U^{id}_{sln} = 48.276 + 0.373x_{Au}, \quad (7.7)$$

$$G^{id}_{sln} = -82.079 - 2.658x_{Au}, \quad (7.8)$$

$$S^{id}_{sln} = 94.529 + 2.27x_{Au}. \quad (7.9)$$

The calculated straight lines U^{id}_{sln}, x_{Au}, G^{id}_{sln}, x_{Au} for the Ag–Au binary solution are shown in Fig. 7.1 (straight line 1), 7.2 (straight line 1). Curves U_{sln}, x_{Au} (Fig. 7.1, curve 2) and G_{sln}, x_{Au} (Fig. 7.2, curve 2) are also given for the sake of comparison.

7.1.2. Thermodynamic properties of mixing a solution and its components

If the mole fractions of substances A and B in a binary solution are equal to x_A and x_B and the energies of their mixing with a solution $\Delta_{mix}U_B$ and $\Delta_{mix}U_A$ at the given concentration are known, the energy of mixing the solution is determined according to the additive equation:

$$\Delta_{mix}U = x_A\Delta_{mix}U_A + x_B\Delta_{mix}U_B. \quad (7.10)$$

The free energy and entropy of mixing are determined in a similar way:

$$\Delta_{mix}G = x_A\Delta_{mix}G_A + x_B\Delta_{mix}G_B, \quad (7.11)$$

$$\Delta_{mix}S = x_A\Delta_{mix}S_A + x_B\Delta_{mix}S_B. \quad (7.12)$$

In equations (7.10) – (7.12), $\Delta_{mix}U$, $\Delta_{mix}G$, and $\Delta_{mix}S$ are the energy, free energy, and entropy of mixing a solution respectively, $\Delta_{mix}U_A$ and $\Delta_{mix}U_B$, $\Delta_{mix}G_A$ and $\Delta_{mix}G_B$, $\Delta_{mix}S_A$ and $\Delta_{mix}S_B$ are the energy, free energy, and entropy respectively for the mixing of substances A and B with a solution. The $\Delta_{mix}U,x_B$ curve is practically determined on the base of the listed quantities. Therefore we will consider how the rest of the quantities mentioned above can be determined with the help of the given curve. The $\Delta_{mix}U,x_{Au}$ curve for the Ag–Au solution at 1379 K is displayed in Fig. 7.3 (curve 1).

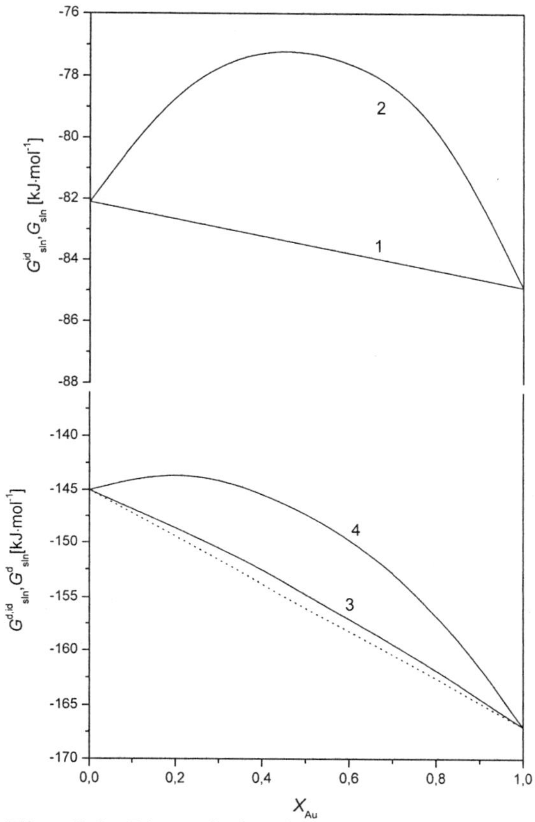

Fig. 7.2. Plot of the free energy of bulk (curves 1 and 2) and nanodispersed (curves 3 and 4) solutions of Ag–Au as a function of x_{Au} at 1379K. Curves: 1– G^{id}_{sln},x_{Au}, 2 – G_{sln},x_{Au}, 3 – $G^{d,id}_{sln},x_{Au}$, 4 – G^{d}_{sln}, x_{Au}.

In order to determine the energies of mixing $\Delta_{mix}U_A$ and $\Delta_{mix}U_B$ for substances A and B with a solution at any value x_B using an experimental curve $\Delta_{mix}U,x_B$ we write equation (7.10) as follows:

$$\Delta_{mix}U = x_A\Delta_{mix}U_A + x_B(\Delta_{mix}U_B - \Delta_{mix}U_A). \quad (7.13)$$

At constant values of $\Delta_{mix}U_A$ and $\Delta_{mix}U_B$ (7.13) is the equation of a straight line. In order to determine magnitudes of $\Delta_{mix}U_A$ and $\Delta_{mix}U_B$ at some value x_B it is therefore necessary to draw the tangent to the $\Delta_{mix}U,x_B$ curve at the given value x_B. The tangent cuts two $\Delta_{mix}U$ axis segments at $x_B = 0$ and $x_B = 1$. According to equation (7.13), their magnitudes are determined by the following equalities:

$$\Delta_{mix}U_{x_B=0} = \Delta_{mix}U_A, \quad (7.14)$$

$$\Delta_{mix}U_{x_B=1} = \Delta_{mix}U_B. \quad (7.15)$$

In order to find values of the energies of mixing $\Delta_{mix}U_{Ag}$ and $\Delta_{mix}U_{Au}$ for the Ag–Au solution at $x_B = x_{Au} = 0.4$, we draw the tangent to the $\Delta_{mix}U,x_{Au}$ curve at the point mentioned (see Fig. 7.3, curve 1). It cuts the following values on the

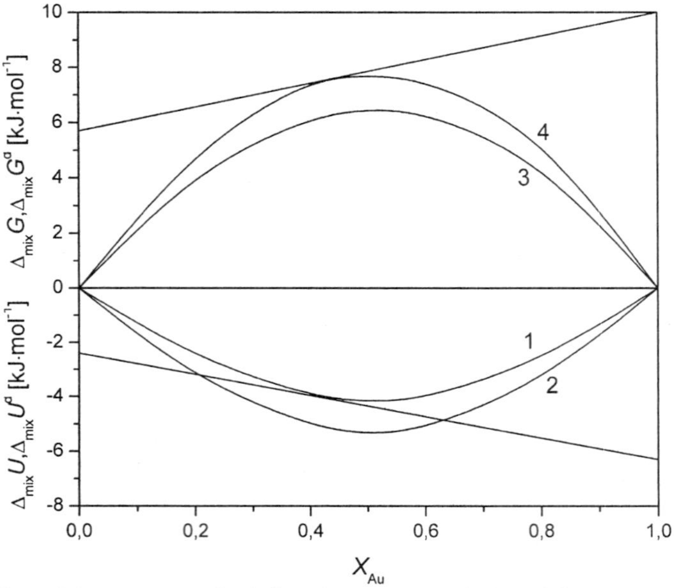

Fig. 7.3. Plot of the energy of mixing (curves 1 and 2) and free energy of mixing (curves 3 and 4) for bulk (curves 1 and 3) and nanodispersed (curves 2 and 4) solutions of Ag–Au as a function of x_{Au} at 1379K. Curves: 1 – $\Delta_{mix}U, x_{Au}$, 2 – $\Delta_{mix}U^d, x_{Au}$, 3 – $\Delta_{mix}G$, x_{Au}, 4 – $\Delta_{mix}G^d, x_{Au}$.

$\Delta_{mix}U$ axis: $\Delta_{mix}U_{x_{Au}=0} = \Delta_{mix}U_{Ag} = -2.45$ kJ·mol^{-1}, $\Delta_{mix}U_{x_{Au}=1} = \Delta_{mix}U_{Au} = -6.20$ kJ·mol^{-1}. On repeating the same operations for other values x_{Au} we have the $\Delta_{mix}U_{Ag}, x_{Au}$ and $\Delta_{mix}U_{Au}, x_{Au}$ curves for the Ag–Au solution.

Let us calculate the free energies of mixing for silver and gold with the Ag–Au solution. For this purpose we shall use the following method [8]. In the defining equations:

$$\Delta_{mix}U_A = U_{A,sln} - U_A, \qquad (7.16)$$

$$\Delta_{mix}U_B = U_{B,sln} - U_B \qquad (7.17)$$

quantities $U_{A,sln}$ and $U_{B,sln}$ can be presented in the following way:

$$U_{A,sln} = \alpha_A U_A, \qquad (7.18)$$

$$U_{B,sln} = \alpha_B U_B. \qquad (7.19)$$

In equations (7.16) – (7.19) U_A, U_B, $U_{A,sln}$, $U_{B,sln}$ are the internal energies of substances A and B in the pure state and in a solution, α_A and α_B are the coefficients determined with the help of experimental data. After substituting (7.18) and (7.19) to (7.16) and (7.17) we obtain the following equations, which allow for calculating coefficients α_A and α_B by experimental values $\Delta_{mix}U_A$ and $\Delta_{mix}U_B$:

$$\Delta_{mix}U_A = (\alpha_A - 1)U_A, \tag{7.20}$$

$$\Delta_{mix}U_B = (\alpha_B - 1)U_B, \tag{7.21}$$

where the internal energies U_A and U_B of pure substances A and B are found using reference data [20–25] (quantities of U_A and U_B should be counted off from $U_0 = 0$ at $T \to 0$).

According to the equation:

$$U = G + TS \tag{7.22}$$

the internal, free energy, and entropy of a substance are interconnected. Therefore the free energies and entropies of mixing the components of a binary solution can be presented by analogy with energies of mixing as follows:

$$\Delta_{mix}G_A = (\alpha_A - 1)G_A, \tag{7.23}$$

$$\Delta_{mix}G_B = (\alpha_B - 1)G_B, \tag{7.24}$$

$$\Delta_{mix}S_A = (\alpha_A - 1)S_A, \tag{7.25}$$

$$\Delta_{mix}S_B = (\alpha_B - 1)S_B. \tag{7.26}$$

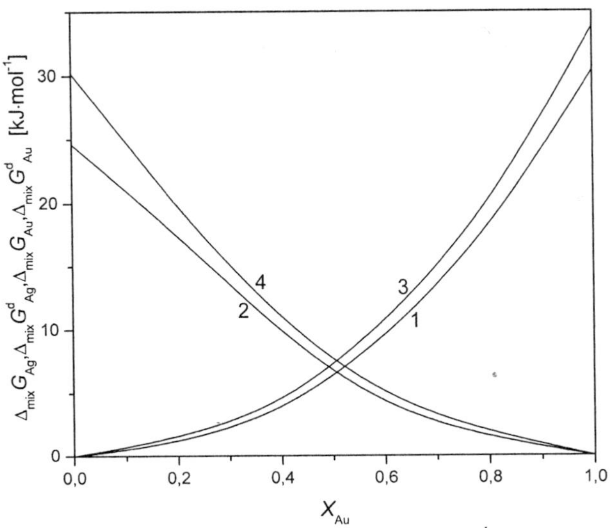

Fig. 7.4. Curves $\Delta_{mix}G_{Ag}, x_{Au}$(1), $\Delta_{mix}G_{Au}, x_{Au}$(2), $\Delta_{mix}G^d_{Ag}, x_{Au}$(3) and $\Delta_{mix}G^d_{Au}$, x_{Au}(4) for the Ag–Au solution at 1379K.

Thus, the calculation of free energies and entropies of mixing the components of the Ag–Au binary solutions comes to finding quantities of (α_A – 1) and (α_B – 1) at various values x_B [8] on equations (7.20) and (7.21) using the found values of $\Delta_{mix}U_{Ag}$ and $\Delta_{mix}U_{Au}$. The $\Delta_{mix}G_{Ag}$,x_{Au} and $\Delta_{mix}G_{Au}$,x_{Au} curves constructed using the values of (α_{Ag} – 1) and (α_{Au} – 1) obtained for the Ag–Au binary solution at various values x_{Au} and equations (7.23) and (7.24) are presented in Fig. 7.4 (curves 1 and 2).

The entropies of mixing silver and gold with the Ag–Au solution can be calculated with the help of known values $\Delta_{mix}U_{Ag}$, $\Delta_{mix}U_{Au}$, $\Delta_{mix}G_{Ag}$ and $\Delta_{mix}G_{Au}$ and the equations:

$$\Delta_{mix}G_A = \Delta_{mix}U_A - T\Delta_{mix}S_A, \qquad (7.27)$$

$$\Delta_{mix}G_B = \Delta_{mix}U_B - T\Delta_{mix}S_B. \qquad (7.28)$$

7.2. Thermodynamic properties of binary nanodispersed solution

7.2.1. Isotherm of the specific surface energy of ideal binary solution

The isotherm of the specific surface energy σ^{id}_{sln} of an ideal binary solution is described by the equation of a straight line [8]:

$$\sigma^{id}_{sln} = \sigma_A + x^{id}_B(sur)(\sigma_A - \sigma_B), \qquad (7.29)$$

where $x^{id}_B(sur)$ is the mole fraction of the component B in the surface of an ideal solution, σ_A and σ_B are the specific surface energies of the pure substances A and B. The quantities $x^{id}_B(sur)$, σ^{id}_{sln}, σ_A and σ_B are interrelated by the equations [8]:

$$\frac{1 - x^{id}_B(sur)}{1 - x_B} = \frac{\sigma^{id}}{\sigma_A} = a, \qquad (7.30)$$

$$\frac{x_B(sur)}{x_B} = \frac{\sigma^{id}}{\sigma_B} = b, \qquad (7.31)$$

where x_B is the mole fraction of substance B in the volume of a solution.

Solving equations (7.30) and (7.31) simultaneously we obtain the expressions for x_B and $x_B(sur)$:

$$x_B = \frac{1-a}{b-a}, \qquad (7.32)$$

$$x_B(sur) = \frac{(1-a)b}{b-a} = x_B b. \qquad (7.33)$$

Setting various values of σ^{id}_{sln}, lying between σ_A and σ_B for each chosen value σ^{id}_{sln} one can find values x_B and $x_B(sur)$ using equations (7.32) and (7.33). The data for the Ag–Au solution are listed in Table 7.1. Experimental values of σ_{sln} [J·m^{-2}] are also displayed.

The σ^{id}_{sln},$x_{Au}(sur)$ and σ^{id}_{sln}, x_{Au} dependences constructed with the help these data are shown in Fig. 7.5 (straight line 1 and curve 2).

Experimental curve σ_{sln}, x_{Au} for the indicated system is shown in Fig. 7.6 (curve 2) and so is the

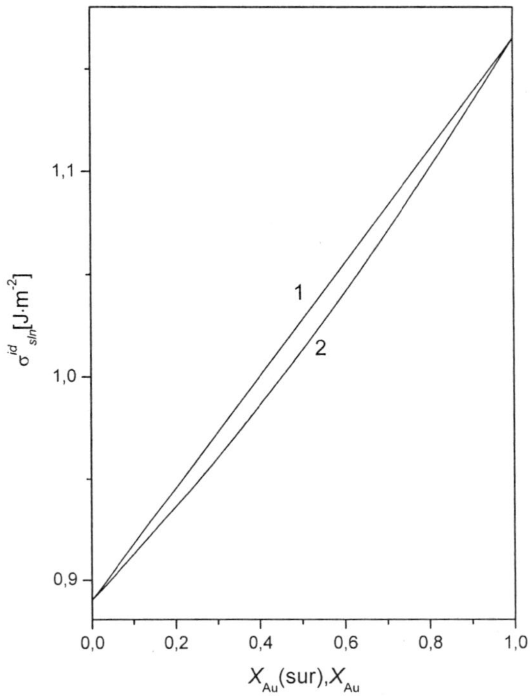

Fig. 7.5. Plot of the specific surface energy of the Ag–Au solution at ideal state at 1379K as a function of $x_A(sur)$ (straight line 1) and x_{Au} (curve 2).

Table 7.1. Values of x_{Au} and $x_{Au}(sur)$ for the Ag–Au solution at 1379 K at various values of σ^{id}_{sln} [J·m^{-2}].

σ^{id}_{sln}	x_{Au}	$x_{Au}(sur)$	σ_{sln}
0.892	0	0	0.892
0.920	0.1295	0.1022	0.912
0.950	0.2598	0.2117	0.937
0.980	0.3820	0.3211	0.965
1.010	0.4970	0.4304	0.993
1.040	0.6055	0.5400	1.023
1.070	0.7079	0.6496	1.055
1.100	0.8047	0.7591	1.089
1.130	0.8965	0.8688	1.123
1.166	1.0	1.0	1.166

$\sigma^{id}_{sln}, x_{Au}$ curve for the ideal solution Ag–Au (curve 1). Some negative deviations from ideality being observed in the Ag–Au solution, the σ_{sln}, x_{Au} curve is below the $\sigma^{id}_{sln}, x_{Au}$ (compare curves 1 and 2 in fig. 7.6).

7.2.2. Thermodynamic properties of ideal nanodispersed solution

In ideal nanodispersed (as well as bulk) solutions thermodynamic properties of mixing ($\Delta_{mix}U^d$, $\Delta_{mix}G^d$ and $\Delta_{mix}S^d$) are equal to zero. Consequently, the internal energy $U^{d,id}_{sln}$, the free energy $G^{d,id}_{sln}$, and the entropy $S^{d,id}_{sln}$ for an ideal nanodispersed solution are equal:

$$U^{d,id}_{sln} = U^{id}_{sln} + U^{S,id}_{sln}, \tag{7.34}$$

$$G^{d,id}_{sln} = G^{id}_{sln} + G^{S,id}_{sln}, \tag{7.35}$$

$$S^{d,id}_{sln} = S^{id}_{sln} + S^{S,id}_{sln}. \tag{7.36}$$

In these equations U^{id}_{sln}, G^{id}_{sln} and S^{id}_{sln} are calculated by equations (7.4) – (7.6). $U^{S,id}_{sln}$, $G^{S,id}_{sln}$, $S^{S,id}_{sln}$ are the surface energy, the free energy, and the entropy of an ideal solution respectively. They depend on the specific surface energy σ^{id}_{sln} or the specific surface entropy s^{S}_{sln} of a solution at the set concentration x_B of an ideal solution and its surface area A:

$$U^{d,id}_{sln} = \sigma^{id}_{sln} A, \tag{7.37}$$

$$G^{d,id}_{sln} = -\sigma^{id}_{sln} A, \tag{7.38}$$

$$S^{d,id}_{sln} = s^{S,id}_{sln} A. \tag{7.39}$$

The quantity $s^{S,id}_{sln}$ for solutions being as a rule unknown, we shall consider the $U^{d,id}_{sln}, x_B$ and $G^{d,id}_{sln}, x_B$ curves for a binary ideal solution. Taking into account (7.4) and (7.5), (7.37) and (7.38) and with the help of equations (7.34) and (7.35) we have:

$$U^{d,id}_{sln} = U_A + x_B(U_B - U_A) + \sigma^{id}_{sln} A, \tag{7.40}$$

$$G^{d,id}_{sln} = G_A + x_B(G_B - G_A) - \sigma^{id}_{sln} A. \tag{7.41}$$

The value of σ^{id}_{sln} at set x_B is determined with the help of calculated curve σ^{id}_{sln}, x_B (see the previous section). $\sigma^{id}_{sln} = \sigma_A$ at $x_B = 0$, and $\sigma^{id}_{sln} = \sigma_B$ when $x_B = 1$.

Let us apply equations (7.40) and (7.41) to the Ag–Au binary solution at 1379 K. In this case U_{Ag} = 48.28 and U_{Au} = 48.65 kJ·mol^{-1}, G_{Ag} = –82.08 and G_{Au} = –84.84 kJ·mol^{-1}, σ_{Ag} = 0.892 and σ_{Au} = 1.166 J·m^{-2}. The values of σ^{id}_{sln} have been found with the help of the σ^{id}_{sln},x_{Au} curve shown in Fig. 7.6 (curve 1). The surface area A of a nanodispersed solution was assumed to be equal to the surface area of nanodispersed silver as a drop of radius r = 0.5 nm. As the molar volume of silver V_{Ag} = 11.733·10^{-6} m^3·mol^{-1}, the surface area A is equal to: $A = 3V_{Ag}/r$ = 70398 m^2·mol^{-1}.

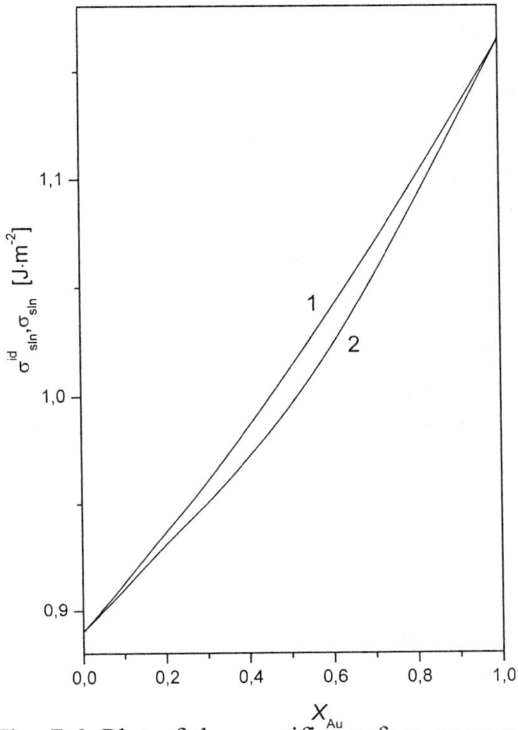

Fig. 7.6. Plot of the specific surface energy of the Ag–Au solution as a function of x_{Au} at ideal (curve 1) and real (curve 2) states at 1379K.

The $U^{d,id}_{sln}$,x_{Au} and $G^{d,id}_{sln}$, x_{Au} curves have been constructed at the above value A using the above mentioned data for Ag and Au, the σ^{id}_{sln},x_{Au} curve (Fig. 7.6, curve 1), and equations (7.40) and (7.41). The curves are shown in Figures 7.1 (curve 3) and 7.2 (curve 3). The $U^{d,id}_{sln}$,x_{Au} and $G^{d,id}_{sln}$,x_{Au} dependencies are nonlinear, by contrast with a bulk ideal solution (compare the straight lines 1 and curves 3 in Figures 7.1 and 7.2). This fact is due to nonlinearity of the σ^{id}_{sln},x_{Au} dependence (see Fig.7.6, curve 1).

7.2.3. Thermodynamic properties of mixing nanodispersed solution and its components

The energy $\Delta_{mix}U^d$ and the free energy $\Delta_{mix}G^d$ of mixing a real nanodispersed solution are made up of quantities $\Delta_{mix}U$ and $\Delta_{mix}G$ of a bulk solution as well as the surface energy $\Delta_{mix}U^S$ and the surface free energy $\Delta_{mix}G^S$ of mixing a solution:

$$\Delta_{mix}U^d = \Delta_{mix}U + \Delta_{mix}U^S, \tag{7.42}$$

$$\Delta_{mix}G^d = \Delta_{mix}G + \Delta_{mix}G^S. \tag{7.43}$$

The quantities $\Delta_{mix}U$ and $\Delta_{mix}G$ are discussed in section 7.1.2. The surface energy $\Delta_{mix}U^S$ and surface free energy $\Delta_{mix}G^S$ of mixing a solution are defined by the equations:

$$\Delta_{mix}U^S = U^d_{sln} - U^{d,id}_{sln}, \tag{7.44}$$

$$\Delta_{mix}G^S = G^d_{sln} - G^{d,id}_{sln}, \tag{7.45}$$

where the quantities $U^{d,id}_{sln}$ and $G^{d,id}_{sln}$ are calculated using equations (7.40) and (7.41), the internal energy U^d_{sln} and the free energy G^d_{sln} of a real nanodispersed solution are calculated by the equations:

$$U^d_{sln} = U_A + x_B(U_B - U_A) + \sigma_{sln}A, \tag{7.46}$$

$$G^d_{sln} = G_A + x_B(G_B - G_A) - \sigma_{sln}A. \tag{7.47}$$

With (7.46), (7.47), (7.40), and (7.41) taking into account, from equations (7.44) and (7.45) follows:

$$\Delta_{mix}U^S = (\sigma_{sln} - \sigma^{id}_{sln})A, \tag{7.48}$$

$$\Delta_{mix}G^S = -(\sigma_{sln} - \sigma^{id}_{sln})A. \tag{7.49}$$

With the help of two equations above as well as equations (7.42) and (7.43) we have:

$$\Delta_{mix}U^d = \Delta_{mix}U + (\sigma_{sln} - \sigma^{id}_{sln})A, \tag{7.50}$$

$$\Delta_{mix}G^d = \Delta_{mix}G - (\sigma_{sln} - \sigma^{id}_{sln})A. \tag{7.51}$$

In the case of the positive deviations from ideal behavior of solution we have $\Delta_{mix}U > 0$ $\sigma_{sln} > \sigma^{id}_{sln}$, $(\sigma_{sln} - \sigma^{id}_{sln}) > 0$. Consequently, $\Delta_{mix}U^d > \Delta_{mix}U$.

This means that the positive deviation from ideality for a substance increases as its dispersity changes from that corresponding to a bulk solution to that in nanodispersed state. For negative deviations from ideality we have $\Delta_{mix}U < 0$, $\sigma_{sln} < \sigma^{id}_{sln}$, $(\sigma_{sln} - \sigma^{id}_{sln}) < 0$. That means that the negative deviations from ideality increase at passage from a bulk solution to nanodispersed one.

Let us apply equations (7.50) and (7.51) to the Ag–Au binary solution at the surface area of A = 70398 $m^2 \cdot mol^{-1}$. For this purpose we find a difference $(\sigma_{sln} - \sigma^{id}_{sln})$ with the help of the σ_{sln},x_{Au} and σ^{id}_{sln},x_{Au} curves (Fig. 7.6, curves 1 and 2) at various values of x_{Au}. Further we calculate the quantities $\Delta_{mix}U^d$ and $\Delta_{mix}G^d$ with the help of the $\Delta_{mix}U$,x_{Au} and $\Delta_{mix}G$,x_{Au} curves and equations (7.50) and (7.51).

The obtained results are shown in Fig. 7.3 (curves 2 and 4). Clearly, negative deviation from ideality increases as the dispersity of a substance changes at passage from the state in the bulk solution Ag–Au to nanodispersed one (Fig. 7.3, curves 1 and 2).

7.2.4. Activity coefficients of the components of nanodispersed solution

The activity coefficient γ_B of substance B in a bulk solution is defined by the equation [8]:

$$\ln \gamma_B = \frac{\Delta_{mix} G_B}{RT}. \tag{7.52}$$

The activity coefficients of silver γ_{Ag} and gold γ_{Au} in the Ag–Au binary solution at 1379K obtained with the help of the $\Delta_{mix}G_{Ag}$,x_{Au} and $\Delta_{mix}G_{Au}$,x_{Au} curves (Fig. 7.4, curves 1 and 2) are displayed in Fig. 7.7 (curves 1 and 2). The activity coefficient γ^d_B of substance B in a nanodispersed solution is defined by the free energy of mixing it with the given solution $\Delta_{mix}G^d_B$:

$$\ln \gamma^d_B = \frac{\Delta_{mix} G^d_B}{RT}. \tag{7.53}$$

We shall use the $\Delta_{mix}G^d$,x_{Au} curve shown in Fig. 7.3 (curve 4) to find the $\Delta_{mix}G^d_{Ag}$, x_{Au} and $\Delta_{mix}G^d_{Au}$,x_{Au} curves for the Ag–Au solution. We shall find the magnitudes of $\Delta_{mix}G_{Au}$ and $\Delta_{mix}G_{Ag}$ at various values of x_{Au} using the tangent method as mentioned above (see section 7.1.2). The tangent to the curve 4 in Fig. 7.3 cuts off two segments $\Delta_{mix}G^d_{Ag} = 5.6$ and $\Delta_{mix}G^d_{Au} = 10$ kJ·mol^{-1} at $x_{Au} = 0.4$ on the $\Delta_{mix}G^d$ axis. The activity coefficients found using equation (7.53) at $A = 70398$ m^2·mol^{-1} are equal to $\ln\gamma^d_{Ag} = 0.488$ and $\ln\gamma^d_{Au} = 0.872$ ($\gamma^d_{Ag} = 1.63$, $\gamma^d_{Au} = 2.39$). The $\ln\gamma^d_{Ag}$,x_{Au} and $\ln\gamma^d_{Au}$,x_{Au} curves obtained in this way are shown in Fig. 7.7 (curves 3 and 4). As is obvious, at the transition from a bulk solution to nanodispersed the activity coefficients of the components of a solution increase ($\gamma^d_{Ag} > \gamma_{Ag}$ and $\gamma^d_{Au} > \gamma_{Au}$). This is due to the fact that the quantity $\Delta_{mix}G^d_B$ develops from $\Delta_{mix}G_B$ and $\Delta_{mix}G^S_B$:

$$\Delta_{mix}G^d_B = \Delta_{mix}G_B + \Delta_{mix}G^S_B. \tag{7.54}$$

The quantity $\Delta_{mix}G^S_B$ is related to the surface activity coefficient γ^S_B of substance B by the equation

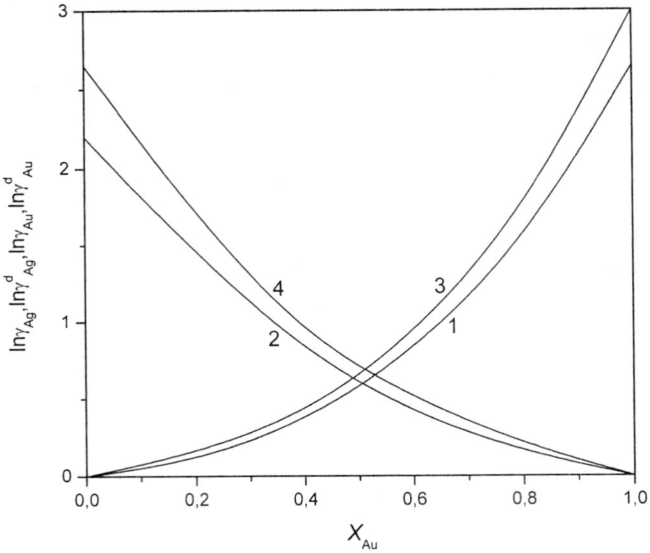

Fig. 7.7. Plot of $\ln\gamma_{Ag}$ (curve 1) and $\ln\gamma_{Au}$ (curve 2), $\ln\gamma^{d}_{Ag}$ (curve 3) and $\ln\gamma^{d}_{Au}$ (curve 4) as a function of x_{Au} for the Ag–Au solution at 1379K.

$$\ln\gamma^{S}_{B} = \frac{\Delta_{mix} G^{S}_{B}}{RT}. \tag{7.55}$$

Taking into account (7.53), (7.54) and (7.55) we obtain the equation

$$\ln\gamma^{d}_{B} = \ln\gamma_{B} + \ln\gamma^{S}_{B} \tag{7.56}$$

or

$$\gamma^{d}_{B} = \gamma_{B} \cdot \gamma^{S}_{B}. \tag{7.57}$$

The value γ^{S}_{B} can be found with the help of known values of γ^{d}_{B} and γ_{B}. Thus for the Ag–Au binary solution we have $\ln\gamma^{d}_{Ag} = 0.92$ and $\ln\gamma_{Ag} = 0.80$ at $x_{Au} = 0.6$. With the help of equation (7.56) we find $\ln\gamma^{S}_{Ag} = 0.12$ ($\gamma^{S}_{Ag} = 1.13$).

7.2.5. Pressure of saturated vapor over nanodispersed solution

For a pure substance the pressures of saturated vapor over nanodispersed and bulk solutions P_S and P^{*}_{S} are related to specific surface energy σ and the surface area A by the equation (see section 5.2)

$$\ln \frac{P_S}{P_S^*} = \frac{\sigma A}{RT}. \tag{7.58}$$

For a solution with specific surface energy σ_{sln} equation (7.58) becomes:

$$\ln \frac{P_{S,sln}}{P_{S,sln}^*} = \frac{\sigma_{sln} A}{RT}. \tag{7.59}$$

Here $P_{S,sln}$ and $P_{S,sln}^*$ are the pressures of saturated vapor over a nanodispersed and a bulk solutions.

If a nanodispersed solution consists of spherical particles of radius r in view of the equation

$$A = \frac{3V_{sln}}{r} \tag{7.60}$$

we find:

$$\ln \frac{P_{S,sln}}{P_{S,sln}^*} = \frac{3\sigma_{sln} V_{sln}}{RTr}. \tag{7.61}$$

Here V_{sln} is the molar volume of a solution, which is calculated for a binary solution by the equation

$$V_{sln} = \frac{M_{sln}}{\rho_{sln}}. \tag{7.62}$$

Here M_{sln}, ρ_{sln} are molar mass and density of a solution. Molar mass of a binary solution is calculated by the equation:

$$M_{sln} = M_A + x_B(M_B - M_A), \tag{7.63}$$

where M_A and M_B are the molar masses of pure substances A and B.

8

THERMODYNAMIC PROPERTIES OF COLLOIDAL DISPERSED SYSTEMS

A nanodispersed substance (liquid or solid) contacting with its vapor (or gas) and liquid in capillaries were treated in the previous chapters. In this chapter, we shall be concerned with the colloidal dispersed systems which consist of a dispersed phase and a liquid named the dispersion medium.

8.1. Aggregative state and the size of dispersed phase particles

The dispersed phase in a liquid dispersion medium most often occurs in solid or liquid state. However gases in the form of bubbles can be found in the liquid state as well. Solid / liquid systems include suspensions, colloidal solutions, while emulsions (for example, milk, medicinal emulsions) pertain to liquid/liquid systems. Finally, gas emulsions may be regarded as gas/liquid systems.

The systems of the dispersed phase/liquid type in which the dispersed phase measures from 1 up to 500 nm are regarded as colloidal dispersed ones. Thus, the colloidal dispersed state of a substance is an intermediate state. It is contiguous with true solutions (the particle sizes is less than 1nm), on the one hand, and with macroscopic substance, on the other hand. Metallic sols (for example, gold sol), biocolloids, macromolecules, micelles, microemulsions are in the colloidal dispersed state.

8.2. Interaction between the dispersed phase and the dispersion medium

As a rule, the dispersion medium is a solution. Even a pure liquid used as the dispersion environment is dissolved, within certain limits, in a liquid. Thus the solution is formed as a result of the contact between the dispersed phase and a pure liquid.

Interaction between the solution and a dispersed phase is mainly reduced to wetting. If the dispersion medium wets a dispersed phase well this system is named lyophilic (in reference to water – hydrophilic). If the dispersion medium does not wet a dispersed phase well, such system is named lyophobic (in reference to water – hydrophobic). Lyophily (hydrophily) or lyophoby (hydrophoby) can be directed by changing the composition of the dispersion medium, provided that the dispersed phase is set.

8.3. The interfacial area of dispersed phase/dispersion medium system

The ratio for two phases of a colloidal dispersed system is frequently expressed in the volume fractions of each phase per liter (dm^3) of the system. Then the volume V_d and the volume fraction x_d of the dispersed phase are equal:

$$V_d = 1 - V_L, \quad (8.1)$$

$$x_d = 1 - x_L, \quad (8.2)$$

where V_L and x_L are the volume of a liquid in one dm^3 of the system and the volume fraction of a liquid.

Let us assume that a particle of a dispersed phase is a sphere of a radius r. Taking then into account that the volume of a particle is $v = 4/3(\pi r^3)$ we have the number of particles of a dispersed phase in one dm^3 of the system (z):

$$z = \frac{V_d}{v} = \frac{3V_d}{4\pi r^3}. \quad (8.3)$$

The surface area of a particle is $a = 4\pi r^2$. Consequently, in view of (8.3) we find the surface area A_d of the dispersed phase in one dm^3 of the system:

$$A_d = za = \frac{3V_d}{r}. \quad (8.4)$$

A_d has dimensionality [$m^2 \cdot dm^{-3}$]. The interfacial area A_{dL} of the dispersed phase/liquid system is equal to the surface area A_d of a dispersed phase: $A_{dL} = A_d$. Assuming that $x_d = 0.4$ and radius of particles of a dispersed phase $r = 5$ nm $= 5 \cdot 10^{-9}$ m, we have $V_d = 0.4 \cdot 10^{-3}\ m^3 \cdot dm^{-3}$ and $A_d = 2.4 \cdot 10^5\ m^2 \cdot dm^{-3}$. If $r = 50 \cdot 10^{-9}$ m, then $A_d = 2.4 \cdot 10^4\ m^2 \cdot dm^{-3}$.

8.4. Surface energy of colloidal dispersed system. Specific interfacial energy.

Both phases in the condensed substance/liquid system possess surface energy contrary to the case of the condensed substance/saturated vapor (gas) system in which only the condensed phase possesses it. However it is only the so-called specific interfacial energy σ_{dL} that can be determined in the system specified.

For the liquid/liquid system one can determine the quantity σ_{dL}, judging from the form of a drop of one liquid (*d*) on the surface of the other one (L). When wetting another substance the drop of a liquid turns to a lens (see Fig. 8.1) with the inclination angle to the horizontal equal to α. Using α on equation [11]:

$$\sigma_{dL} = \left(\sigma_L^2 + \sigma_d^2 - 2\sigma_L\sigma_d\cos\alpha\right)^{0.5} \tag{8.5}$$

specific interfacial energy σ_{dL} can be determined, provided that the specific surface energies of a dispersed phase σ_d and a dispersion liquid σ_L are known.

Surface energy U_{dL}^S of the colloidal dispersed system liquid / liquid depends on the surface area A_d [$m^2 \cdot dm^{-3}$]:

$$U_{dL}^S = \sigma_{dL} A_d. \tag{8.6}$$

Energy U_{dL}^S [$J \cdot dm^{-3}$] needs to be spent to form the emulsion in which the liquid of a disperse phase is distributed in a dispersion liquid as small particles. Thus the free surface energy G_{dL}^S of colloidal dispersed system will decrease by an amount of

$$G_{dL}^S = -\sigma_{dL} A_d. \tag{8.7}$$

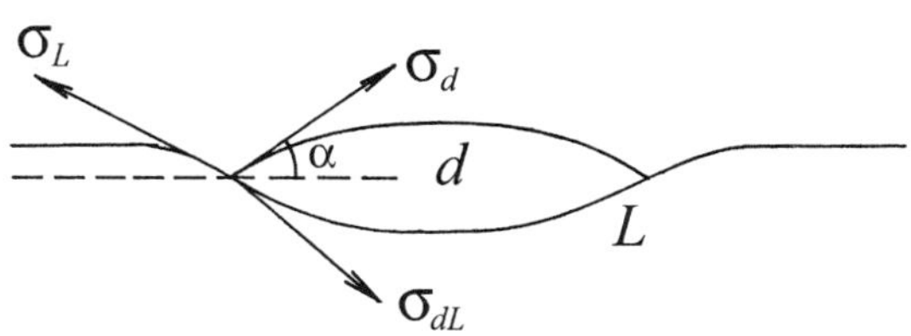

Fig. 8.1. Formation of a lens of liquid *d* on the surface of liquid *L*: σ_L and σ_d are the specific surface energies of liquids *d* and *L*, σ_{dL} is the specific interfacial energy at liquid *d*/liquid *L* interface.

Let us estimate U^S_{dL} for the formation of macro- and microemulsions. The emulsions with the particle size from 1 mm (10^{-3} m) up to 5 mcm ($5\cdot10^{-6}$ m) are referred to macroemulsions, while those with the particle size from $5\cdot10^{-6}$ up to 10^{-8} m to microemulsions. $V_d = 0.4\cdot10^{-3}$ m^3·dm^{-3} when $x_d = 0.4$. On equation (8.4) we find that the interfacial area A_d is equal to $2.4 - 4.8\cdot10^2$ m^2·dm^{-3}, as far as the formation of macroemulsions and to $4.8\cdot10^2 - 2.4\cdot10^5$ m^2·dm^{-3} at the case of microemulsions.

From experience follows that in aqueous solution of surface-active materials a macroemulsion of olive oil that is stable enough can be obtained at a specific interfacial energy σ_{dL}= 0.010 J·m^{-2}. At this value of σ_{dL} the surface energy of macroemulsions ranges from 0.24 to 4.8 J·dm^{-3}.

Microemulsions can be successfully produced if the specific interfacial energy is decreased to 0.0001 J·m^{-2}. At this value σ_{dL} the surface energy of microemulsion is equal to 0.048 – 24 J·dm^{-3} (A_d=$4.8\cdot10^2 - 2.4\cdot10^5$ m^2·dm^{-3}). If σ_{dL} is decreased further down to $2\cdot10^{-6}$ J·m^{-2}, spontaneous emulsification is observed in the system of olive oil/aqueous solution of surface-active materials (tensides). Fig. 8.2 shows the change in the surface energy U^S_{dL} and the surface free energy G^S_{dL} at formation of microemulsion with an increase in the interfacial area A_d at σ_{dL}= 0.0001 J·m^{-2}.

An analysis of the data makes it reasonable to assume that a microemulsion which is stable enough can be obtained by forced mixing the dispersed phase/liquid system, provided that the energy required is approximately 10 J·dm^{-3}.

The precondition for spontaneous emulsification is that the surface energy of formation of emulsion composition should be nearly zero. Therefore we shall formulate the thermodynamic conditions resulting in spontaneous emulsification.

8.5. Thermodynamic conditions for spontaneous formation of colloidal dispersed liquid/liquid mixture

Surface energy U^S_{dL} is an amount of energy that needs to be spent at formation of emulsion with the set interfacial area A_d. Let us consider the thermodynamic characteristics of the liquid / liquid system influencing U^S_{dL}.

According to equations (8.6) and (8.5), at constant value A_d the interfacial energy U^S_{dL} is determined only by the specific interfacial energy σ_{dL}, which depends on following three parameters of the liquid / liquid system (see equation (8.5)):

– specific surface energy σ_L of a solution of tensides;

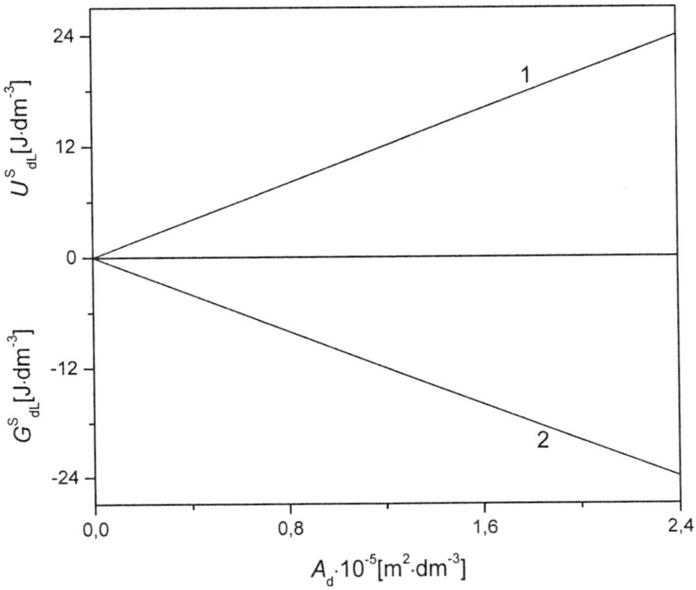

Fig. 8.2. The interfacial (U^S_{dL}) and free interfacial (G^S_{dL}) energies versus the interfacial area A_d when σ_{dL}= 0.0001 J·m^{-2}.

- specific surface energy σ_d of the liquid of a dispersed phase in which tensides are also dissolved;
- inclination angle α of a drop of the liquid in a dispersed phase on the surface of the liquid of the dispersion medium (see Fig. 8.1).

Magnitude σ_L can be reduced by dissolving tensides in liquid L. In this case, the angle α decreases simultaneously. When the angle α is equal to zero equation (8.5) becomes:

$$\sigma_{dL} = \sigma_L - \sigma_d. \tag{8.8}$$

By dissolving tensides in a dispersion liquid at $\alpha = 0$, we reduce its specific surface energy σ_L as much as required for the equality to be true:

$$\sigma_L = \sigma_d. \tag{8.9}$$

In this case σ_{dL}= 0 and U^S_{dL}= 0. Under these conditions, there is a spontaneous mixing of liquid d and liquid L leading to the formation of colloidal dispersed composition.

Thus, the thermodynamic conditions leading to a spontaneously formation of a colloidal dispersed composition can be formulated as follows:

$$\alpha = 0,$$

$$\sigma_L - \sigma_d = 0.$$

In other words, the liquid of a dispersed phase should wet the surface of a liquid of the dispersion medium completely and the specific surface energies of both liquids should be identical.

9

EQUILIBRIUM OF CHEMICAL REACTION INVOLVING NANODISPERSED SUBSTANCES

When the surface area of a condensed substance increases, so do its internal energy and entropy, with its free energy, on the contrary, decreasing. Therefore, if nanosized substances participate in chemical reaction, the standard free energy $\Delta_r G^*$ of reaction undergoes changes and so does the equilibrium constant, as a result. That means that within certain limits, it is possible to govern this chemical transformation by grinding the participating condensed substances. In this chapter the influence of the surface area of the condensed substances on the equilibrium of chemical transformation is discussed. The findings are compared with the data for reactions with participation of bulk substances.

9.1. The law of mass action. The equilibrium constant and the coefficients μ and γ of chemical reaction

If the substances C and D react forming substances M and N when placed in a closed vessel:

$$\nu_C \, \mathrm{C} + \nu_D \, \mathrm{D} = \nu_M \, \mathrm{M} + \nu_N \, \mathrm{N}, \tag{9.1}$$

an equilibrium between the indicated substances will be established at given temperature T and pressure P. The substances in equilibrium have the following concentrations: x_C, x_D, x_M, x_N. Some or all substances can form solute (gaseous, liquid or solid solutions) with the following activity coefficients: γ_C, γ_D, γ_M, and γ_N. The interrelation between the equilibrium concentrations, the activity coefficients, and the relative pressure f of the gaseous participants of a reaction is determined by the law of mass action [8, 9]:

$$K = \frac{\gamma_M^{\nu_M} \cdot \gamma_N^{\nu_N}}{\gamma_C^{\nu_C} \cdot \gamma_D^{\nu_D}} \cdot \frac{x_M^{\nu_M} \cdot x_N^{\nu_N}}{x_C^{\nu_C} \cdot x_D^{\nu_D}} f^{\sum_B \nu_B}. \tag{9.2}$$

In equations (9.1) and (9.2) ν_C, ν_D, ν_M, ν_N are the stoichiometric coefficients, of which the values are determined by the law of conservation of matter. K is the equilibrium constant.

In equation (9.2) the sum $\sum_B \nu_B$ is determined by the values of stoichiometric coefficients of the gaseous participants in reaction (9.1). An example is reaction (9.1) in the gas phase:

$$\sum_B \nu_B = \nu_M + \nu_N - \nu_C - \nu_D. \tag{9.3}$$

If only substances D and N are in the gas phase:

$$\sum_B \nu_B = \nu_N - \nu_D. \tag{9.4}$$

Let us introduce the following designations for the ease of writing:

$$\gamma = \frac{\gamma_M^{\nu_M}\gamma_N^{\nu_N}}{\gamma_C^{\nu_C}\gamma_D^{\nu_D}} = \prod_B \gamma_B^{\nu_B}, \tag{9.5}$$

$$\mu = \frac{x_M^{\nu_M} x_N^{\nu_N}}{x_C^{\nu_C} x_D^{\nu_D}} f^{\sum_B \nu_B} = f^{\sum_B \nu_B} \prod_B x_B^{\nu_B}. \tag{9.6}$$

Then the law of mass action can be rewritten as:

$$K = \gamma\mu . \tag{9.7}$$

Thus, the equilibrium constant is the product of two coefficients γ and μ. The equilibrium constant for given reaction at constant temperature and pressure is constant, therefore a change in any coefficients (γ or μ) leads to a change in other in such a way that their product remains constant.

If only one of participants of chemical reaction is a gas, for example the substance N in reaction (9.1), equations (9.2), (9.5), and (9.6) become:

$$K = \frac{\gamma_M^{\nu_M}}{\gamma_C^{\nu_C}\gamma_D^{\nu_D}} \cdot \frac{x_M^{\nu_M}}{x_C^{\nu_C} x_D^{\nu_D}} f_N, \tag{9.8}$$

$$\gamma = \frac{\gamma_M^{\nu_M}}{\gamma_C^{\nu_C}\gamma_D^{\nu_D}}, \tag{9.9}$$

$$\mu = \frac{x_M^{\nu_M}}{x_C^{\nu_C} x_D^{\nu_D}} f_N. \tag{9.10}$$

The dissociation of an oxide may be taken as an example of reaction with one gas participant:

$$MO = M + 0.5O_2. \tag{9.11}$$

Some examples of application of the law of mass action are listed below.

9.1.1. Reduction of oxide by gas

A metal M can be reduced from oxide (for example, MO) using hydrogen or carbon monoxide:

$$MO + H_2 = M + H_2O. \tag{9.12}$$

In this case, $\sum_B \nu_B = 0$. Consequently, for reaction (9.12) we have the equations

$$K = \frac{\gamma_{M}\gamma_{H_2O}}{\gamma_{MO}\gamma_{H_2}} \cdot \frac{x_{M}x_{H_2O}}{x_{MO}x_{H_2}}, \tag{9.13}$$

$$\gamma = \frac{\gamma_{M}\gamma_{H_2O}}{\gamma_{MO}\gamma_{H_2}}, \tag{9.14}$$

$$\mu = \frac{x_{M}x_{H_2O}}{x_{MO}x_{H_2}} = \frac{x_{M}f_{H_2O}}{x_{MO}f_{H_2}}. \tag{9.15}$$

9.1.2. Reactions between the gas phase and condensed substances

Let us consider the Boudoir's reaction as an example:

$$C + CO_2 = 2CO\,, \tag{9.16}$$

for which $\sum_B \nu_B = 2 - 1 = 1$. Therefore:

$$K = \frac{\gamma_{CO}^2}{\gamma_{CO_2}\gamma_{C}} \cdot \frac{x_{CO}^2\, x_{C}}{x_{CO_2}} f, \tag{9.17}$$

$$\gamma = \frac{\gamma_{CO}^2}{\gamma_{CO_2}\gamma_{C}}, \tag{9.18}$$

$$\mu = \frac{x_{CO}^2\, x_{C}}{x_{CO_2}} f = \frac{x_{C}f_{CO}^2}{f_{CO_2}}. \tag{9.19}$$

9.1.3. Decomposition of condensed substance with formation of gas

The decomposition of calcium carbonate is an example of such a reaction:

$$CaCO_3 = CaO + CO_2 \,. \tag{9.20}$$

In the case equations for the equilibrium constant and the coefficients of γ and μ become:

$$K = \frac{\gamma_{\text{CaO}}}{\gamma_{\text{CaCO}_3}} \cdot \frac{x_{\text{CaCO}_3}}{x_{\text{CaO}}} f_{\text{CO}_2}, \tag{9.21}$$

$$\gamma = \frac{\gamma_{\text{CaO}}}{\gamma_{\text{CaCO}_3}}, \tag{9.22}$$

$$\mu = \frac{x_{\text{CaCO}_3}}{x_{\text{CaO}}} f_{\text{CO}_2}\,. \tag{9.23}$$

9.2. Thermodynamic characteristics of chemical reaction

The changes in energy at chemical reaction are caused by the following processes:

- New substances are formed as the products of reaction;
- The participants of chemical reaction pass into a solution (gas mixture).

These processes run simultaneously in an actual chemical reaction. However, they can be considered in turn, for clarity sake.

The formation of reaction products from parent materials can be presented as the standard reaction:

$$\nu_C \text{C}^* + \nu_D \text{D}^* = \nu_M \text{M}^* + \nu_N \text{N}^*. \tag{9.24}$$

The process of substances C, D, M, and N passing into a solution can be presented in the general form:

$$\text{B}^* = \text{B}. \tag{9.25}$$

Combining the equation of the normal reaction (9.24) and equation (9.25) for each participant of the reaction passing into a solution with the corresponding signs (plus for substances M and N, minus for substances C and D) we obtain the equation of the reaction (9.1):

$$\nu_C\,\mathrm{C}+\nu_D\,\mathrm{D}=\nu_M\,\mathrm{M}+\nu_N\,\mathrm{N}\,. \tag{9.1}$$

The reaction (9.24) is characterized by the following standard thermodynamic characteristics: the energy $\Delta_r U^*$, the free energy $\Delta_r G^*$, and the entropy $\Delta_r S^*$, while the passage of the participants of the reaction into a solution is described by the following thermodynamic characteristics of mixing: the energy $\Delta_r\Delta_{mix}U$, the free energy $\Delta_r\Delta_{mix}G$, and the entropy $\Delta_r\Delta_{mix}S$. From the aforesaid follows that:

$$\Delta_r U=\Delta_r U^*+\Delta_r\Delta_{mix}U, \tag{9.26}$$

$$\Delta_r G=\Delta_r G^*+\Delta_r\Delta_{mix}G, \tag{9.27}$$

$$\Delta_r S=\Delta_r S^*+\Delta_r\Delta_{mix}S, \tag{9.28}$$

where $\Delta_r U$ is the energy, $\Delta_r G$ is the free energy, and $\Delta_r S$ is the entropy of reaction (9.1) respectively.

Thus, the chemical reaction (9.1) is characterized by the following thermodynamic characteristics:
energy $\Delta_r U$,
standard energy $\Delta_r U^*$,
energy of mixing $\Delta_r\Delta_{mix}U$,
free energy $\Delta_r G$,
standard free energy $\Delta_r G^*$,
free energy of mixing $\Delta_r\Delta_{mix}G$,
entropy $\Delta_r S$,
standard entropy $\Delta_r S^*$,
entropy of mixing $\Delta_r\Delta_{mix}S$.

The thermodynamic characteristics listed above are defined as follows:

$$\Delta_r U^*=\nu_M U_M^*+\nu_N U_N^*-\nu_C U_C^*-\nu_D U_D^* \tag{9.29}$$

or $\Delta_r U^*=\sum_B \nu_B U_B^*$,

$$\Delta_r G^*=\nu_M G_M^*+\nu_N G_N^*-\nu_C G_C^*-\nu_D G_D^* \tag{9.30}$$

or $\Delta_r G^*=\sum_B \nu_B G_B^*$,

$$\Delta_r S^*=\nu_M S_M^*+\nu_N S_N^*-\nu_C S_C^*-\nu_D S_D^* \tag{9.31}$$

or $\Delta_r S^*=\sum_B \nu_B S_B^*$.

By analogy with the foregoing we have for other quantities:

$$\Delta_r U = \sum_B \nu_B U_B, \tag{9.32}$$

$$\Delta_r G = \sum_B \nu_B G_B, \tag{9.33}$$

$$\Delta_r S = \sum_B \nu_B S_B, \tag{9.34}$$

$$\Delta_r \Delta_{mix} U = \sum_B \nu_B \Delta_{mix} U_B, \tag{9.35}$$

$$\Delta_r \Delta_{mix} G = \sum_B \nu_B \Delta_{mix} G_B, \tag{9.36}$$

$$\Delta_r \Delta_{mix} S = \sum_B \nu_B \Delta_{mix} S_B. \tag{9.37}$$

In these equations U_B^* and U_B, G_B^* and G_B, S_B^* and S_B represent the energy, the free energy, and the entropy for any of participants of the reaction in the pure state and in a solution respectively, while $\Delta_{mix}U_B$, $\Delta_{mix}G_B$ and $\Delta_{mix}S_B$ express the energy, the free energy, and the entropy of mixing for any of the participants of the reaction with a solution.

The mentioned thermodynamic characteristics of the reaction are interrelated by the following equations:

$$\Delta_r U = \Delta_r G + T\Delta_r S, \tag{9.38}$$

$$\Delta_r U^* = \Delta_r G^* + T\Delta_r S^*, \tag{9.39}$$

$$\Delta_r \Delta_{mix} U = \Delta_r \Delta_{mix} G + \Delta_r \Delta_{mix} S. \tag{9.40}$$

The equations listed in sections 9.1 and 9.2 are valid for variable temperature at any pressure, provided it is constant.

9.3. Interrelation between the thermodynamic characteristics of a chemical reaction, the equilibrium constant, and coefficients μ and γ

The equilibrium constant K is related to the standard free energy $\Delta_r G^*$ of a reaction by the equation

$$RT\ln K = -\Delta_r G^* . \tag{9.41}$$

If $\Delta_r\Delta_{mix}G = 0$, the defining equation for the coefficient μ follows from equations (9.27) and (9.41):

$$RT\ln\mu = -\Delta_r G . \tag{9.42}$$

The coefficient γ can be determined by equation (9.7), if the values of K and μ are known:

$$\gamma = \frac{K}{\mu} . \tag{9.43}$$

Taking the logarithm of equation (9.43) we have:

$$\ln\gamma = \ln K - \ln\mu \tag{9.44}$$

or

$$RT\ln\gamma = RT\ln K - RT\ln\mu . \tag{9.45}$$

Taking into account equations (9.27), (9.41), and (9.42) we come to the defining equation for the coefficient γ:

$$RT\ln\gamma = \Delta_r\Delta_{mix}G . \tag{9.46}$$

According to equations (9.5) and (9.36) we have for any participant of a reaction in a solution:

$$RT\ln\gamma_B = \Delta_{mix}G_B = G_B - G_B^* , \tag{9.47}$$

where γ_B is the activity coefficient, G_B and G_B^* are the free energies of a substance B in a solution and in the pure state, $\Delta_{mix}G_B$ is the free energy of mixing of a substance B with a solution.

The equations listed in the current section are valid for bulk condensed substances at any temperature and at any constant pressure.

9.4. Standard thermodynamic characteristics of a chemical reaction involving nanodispersed substances

If the condensed participants of a chemical reaction are nanosized particles, their thermodynamic properties (U_B^*, G_B^*, S_B^*) depend on the surface area, not just on temperature and pressure. For any pure substance B the U_B^*, G_B^*, S_B^* dependence on the surface area is defined by the equations (see chapter 3)

$$U_B^{*,d} = U_B^* + \sigma_B A_B, \tag{9.48}$$

$$G_B^{*,d} = G_B^* - \sigma_B A_B, \tag{9.49}$$

$$S_B^{*,d} = S_B^* + s_B A_B, \tag{9.50}$$

where $U_B^{*,d}$ and U_B^*, $G_B^{*,d}$ and G_B^*, $S_B^{*,d}$ and S_B^* are respectively the internal energies, free energies, and entropies of nano- and macrophases of substance B, σ_B and A_B are the specific surface energy of substance B at the border with gas and its molar surface area respectively, s_B is the specific surface entropy of substance B.

The quantity A_B depends on the form of nanoparticles. For spherical particles of radius r

$$A_B = \frac{3V_B}{r}, \tag{9.51}$$

while for threadlike particles with a string radius of r

$$A_B = \frac{2V_B}{r}, \tag{9.52}$$

where V_B is the molar volume of substance B.

Let us assume that the participants C and M are condensed substances. Then quantities $\Delta_r U^{*,d}$, $\Delta_r G^{*,d}$ and $\Delta_r S^{*,d}$ are defined in the following way:

$$\Delta_r U^{*,d} = \sum_B \nu_B U_B^* + \sum_B \nu_B \sigma_B A_B, \tag{9.53}$$

$$\Delta_r G^{*,d} = \sum_B \nu_B G_B^* - \sum_B \nu_B \sigma_B A_B, \tag{9.54}$$

$$\Delta_r S^{*,d} = \sum_B \nu_B S_B^* + \sum_B \nu_B s_B A_B. \tag{9.55}$$

The quantities $\sum_B \nu_B U_B^*$, $\sum_B \nu_B G_B^*$, and $\sum_B \nu_B S_B^*$ are calculated according to equations (9.29) – (9.31), and the rest quantities are defined by the equations

$$\sum_B \nu_B \sigma_B A_B = \nu_C \sigma_C A_C - \nu_M \sigma_M A_M, \tag{9.56}$$

$$\sum_B \nu_B s_B A_B = \nu_C s_C A_C - \nu_M s_M A_M, \tag{9.57}$$

where s_C, s_M and A_C, A_M are the specific surface entropy and the molar surface area of substances C and M. It is necessary to keep in mind that only the condensed substances participating in a chemical transformation are taken into account in equations (9.56) and (9.57). Equation (9.56) suggests that the magnitude $\sum_B \nu_B \sigma_B A_B$ can be both positive and negative. If $\sum_B \nu_B \sigma_B A_B > 0$ from equations (9.53) – (9.55) follows that $\Delta_r U^{*,d} > \Delta_r U^*$, $\Delta_r G^{*,d} < \Delta_r G^*$, and $\Delta_r S^{*,d} > \Delta_r S^*$. When $\sum_B \nu_B \sigma_B A_B < 0$ then $\Delta_r U^{*,d} < \Delta_r U^*$, $\Delta_r G^{*,d} > \Delta_r G^*$, and $\Delta_r S^{*,d} < \Delta_r S^*$.

The foresaid implies that with an increase in the surface area of the condensed substances equilibrium of a chemical reaction (9.1) can shift both to the right and to the left.

9.5. Equilibrium constant of chemical reaction involving nanodispersed substances

Equilibrium constants K and K^d of the chemical reaction (9.1) with participation of bulk and nanodispersed substances are related with standard free energies $\Delta_r G^*$ and $\Delta_r G^{*,d}$ by the equations

$$\ln K = -\frac{\Delta_r G^*}{RT}, \tag{9.58}$$

$$\ln K^d = -\frac{\Delta_r G^{*,d}}{RT}, \tag{9.59}$$

which allow us to write the following equation:

$$\ln \frac{K^d}{K} = -\frac{1}{RT}\left(\Delta_r G^{*,d} - \Delta_r G^*\right). \tag{9.60}$$

Designating

$$\Delta_r G^S = \Delta_r G^{*,d} - \Delta_r G^* \tag{9.61}$$

we have

$$\ln \frac{K^d}{K} = -\frac{\Delta_r G^{*,S}}{RT}, \tag{9.62}$$

where the standard free surface energy $\Delta_r G^{*,S}$ is calculated according to

$$\Delta_r G^{*,S} = -\sum \nu_B \sigma_B A_B. \tag{9.63}$$

The condensed participants of a chemical reaction (see (9.56)) are taken into account in the equation only.

From equation (9.62) follows that $K^d > K$, if $\Delta_r G^{*,S} < 0$ and $K^d < K$ if $\Delta_r G^{*,S} > 0$, that is in the first case equilibrium shifts to the right, and in the second case to the left.

It is worth reminding that the surface area A_B for spherical particles is calculated by equation (9.51).

Let us discuss some examples of the K^d/K ratio being calculated, assuming that the radius of the spherical particles participating in reactions changes from 50 up to 0.5 nm.

For the formation of the liquid spherical particles of silicon dioxide by oxidation of liquid silicon particles we have

$$Si + O_2 = SiO_2. \tag{9.64}$$

The data necessary for calculation $\Delta_r G^{*,S}$ at 2023 K are listed below:

Substance	V [$m^3 \cdot mol^{-1}$]	σ_B [$J \cdot m^{-2}$]
Si	$12.39.10^{-6}$	0.679
SiO_2	$28.57.10^{-6}$	0.400

For pure bulk substances equilibrium constant of reaction (9.64) is equal to

$$K = P_{O_2}^{-1}. \tag{9.65}$$

$\Delta_r G^* = -537801$ $J \cdot mol^{-1}$ [23] when $T = 2023$ K. Taking into account the equation

$$\ln K = -\frac{\Delta_r G^*}{RT}, \tag{9.66}$$

we find $K = 7.70 \cdot 10^{13}$. In other words, equilibrium of reaction (9.64) undergoes nearly complete shift to the right. At the same time, equilibrium pressure of oxygen $P_{O_2} = 1.23 \cdot 10^{-12}$ bar at $T = 2023$K.

When Si is oxidized to SiO_2 the volume increases by a factor of 2.306. Consequently, the radius of formed spherical particle of SiO_2 increases by a factor of 1.32. Using the data listed above and taking into the account that $r_{SiO_2} = 1.32 r_{Si}$ we shall calculate $\Delta_r G^{*,S}$ and the K^d/K ratio at $r_{Si} = 50$, 5, and 0.5 nm, i.e. $r_{SiO_2} = 66$, 6.6, and 0.66 nm. The results are shown below:

Substance, reaction	Values A [$m^2 \cdot mol^{-1}$], G^S_{Si}, $G^S_{SiO_2}$, $\Delta_r G^{*,S}$ [$J \cdot mol^{-1}$] and K^d/K at r_{Si} [m], equal to		
	$50 \cdot 10^{-9}$	$5 \cdot 10^{-9}$	$0.5 \cdot 10^{-9}$
		A	
Si	743	7430	74300
SiO_2	1295	1290	12900
		G^S_{Si}, $G^S_{SiO_2}$	
Si	-504.5	-5045	-50450
SiO_2	-518	-5180	-51800
		$\Delta_r G^{*,S}$	
reaction (9.64)	-13.5	-135	-1350
		K^d/K	
reaction (9.64)	1.0008	1.008	1.083

From the data above follows that an increase in the surface area of silicon has practically no influence on the equilibrium constant of reaction (9.64).

For the oxidation of liquid drops of iron

$$Fe + 0.5O_2 = FeO \qquad (9.67)$$

at 1808 K the specific surface energies and molar volumes of Fe and FeO equal:

Substance	V [$m^3 \cdot mol^{-1}$]	σ_B [$J \cdot m^{-2}$]
Fe	$7.9 \cdot 10^{-6}$	1.86
FeO	$12.60 \cdot 10^{-6}$	0.59

The standard free energy and equilibrium constant of reaction (9.67) are equal: $\Delta_r G^* = -143308$ $J \cdot mol^{-1}$, $K = 1.38 \cdot 10^4$, i.e. equilibrium of reaction (9.67) is significantly shifted to the right. Equilibrium pressure of oxygen $P_{O_2} = 5.25 \cdot 10^{-7}$ bar.

The derived values $\Delta_r G^{*,S}$, K^d/K and P_{O_2} for reaction (9.67) are listed below at $r_{Fe} = 50$, 5, and 0.5 nm with the account that $r_{FeO} = 1.17\ r_{Fe}$.

Substance, reaction	Values A [$m^2 \cdot mol^{-1}$], G^S_{Fe}, G^S_{FeO}, $\Delta_r G^{*,S}$ [$J \cdot mol^{-1}$] and K^d/K, K^d and P_{O_2} [bar] at r_{Fe} [m], equal to:		
	$50 \cdot 10^{-9}$	$5 \cdot 10^{-9}$	$0.5 \cdot 10^{-9}$
	A		
Fe	474.6	4746	47460
FeO	756	7560	75600
	G^S_{Fe}, G^S_{FeO}		
Fe	-882.7	-8827	-88270
FeO	-446	-4460	-44600

		$\Delta_r G^{*,S}$		
reaction (9.67)	436.7		4367	43670
		K^d/K		
reaction (9.67)	0.971		0.748	0.0547
		K^d		
reaction (9.67)	$1.34 \cdot 10^4$		$1.03 \cdot 10^4$	$7.55 \cdot 10^2$
		P_{O_2}		
reaction (9.67)	$5.58 \cdot 10^{-7}$		$9.42 \cdot 10^{-7}$	$1.75 \cdot 10^{-2}$

An analysis of the data reveals that as state of iron changes from bulk to spherical particles with the radius of 0.5 nm equilibrium constant of reaction (9.67) decreases from $1.34 \cdot 10^4$ to $7.55 \cdot 10^2$, i.e. by a factor of 18. At the same time, the equilibrium pressure of oxygen calculated by the equation

$$P_{O_2} = (K^d)^{-2} \tag{9.68}$$

increases from $5.25 \cdot 10^{-7}$ to $1.75 \cdot 10^{-2}$ bar, i.e. by a factor of $1.9 \cdot 10^4$. This means, that in the atmosphere of oxygen iron nanoparticles start to be oxidized at more high pressure than the macrophase.

If the gas phase consists of H_2S and SO_2, the reaction

$$2H_2S + SO_2 = S + 2H_2O \tag{9.69}$$

can lead to the generation of solid sulfur. The standard free energy of the reaction at 298K $\Delta_r G^* = -90483$ J·mol^{-1}, and equilibrium constant $K = 7.258 \cdot 10^{15}$.

Assuming that the radius of spherical particles of sulfur is equal to 50, 5, and 0.5 nm and, taking into account $V = 16.36 \cdot 10^{-6}$ m^2·mol^{-1} and $\sigma = 0.066$ J·m^{-2}, we find $\Delta_r G^{*,d} = -90677, -92426$, and -109918 J·mol^{-1} and $K^d = 7.849 \cdot 10^{15}$, $1.590 \cdot 10^{16}$, and $1.851 \cdot 10^{19}$. In other words, if the radius of the nanoparticles of sulfur decreases from 50 to 0.5 nm, the equilibrium constant of reaction (9.69) increases by a factor of $2.36 \cdot 10^3$, i.e. more than by three orders of magnitude.

10

FROM THE ENERGY OF ATOMS IN SOLID TO THE ENERGY OF ATOMS IN GAS

When a solid turns into the gas phase its internal energy raises by an amount equal to the energy of sublimation $\Delta_{sub}U$ which can be found in reference books. The energy of sublimation of nanodispersed substance decreases, its internal energy being higher than that of bulk substance. This chapter treats of the change in energy of substance when passing from bulk to nanodispersed state and from nanodispersed to gaseous (nuclear or molecular) state.

10.1. Energy of bulk substance

When the thermodynamic characteristics of bulk substance are defined its surface energy is neglected, since the part of atoms or molecules at the surface does not normally exceed 10^{-7}. For example, the surface area of one mole of gold as the sphere with a volume of $10.215\cdot10^{-6}$ m$^3\cdot$mol^{-1} is equal to $2.277\cdot10^{-3}$ m$^2\cdot$mol^{-1}. Over this area $3.16\cdot10^{16}$ atoms of gold can be allocated. Hence follows that the part of atoms at the surface is equal to $0.525\cdot10^{-7}$ and can be neglected.

Let us find the part of surface energy of gold at 298K. At this temperature the internal energy of gold is equal to 5680 J$\cdot$mol^{-1} [23, 27]. The specific surface energy of gold at 298K equals 1.307 J$\cdot$m^{-2} (see section 4.2). Consequently, its surface energy is equal: $U^S = 1.307\cdot2.277\cdot10^{-3} = 2.976\cdot10^{-3}$ J$\cdot$mol^{-1}. It follows that the part of surface energy makes $5.34\cdot10^{-7}$. In other words, the surface energy of bulk substance can be neglected.

10.2. From the energy of bulk substance to the energy of nanodispersed substance

In nanodispersed state a diameter of one spherical particle decreases from 100 to 1 nanometer. At the same time the molar surface area can be calculated according to the equation

$$A = \frac{3V}{r}, \tag{10.1}$$

where V is the molar volume of a substance. For gold at 298K $V = 10.215 \cdot 10^{-6}$ $m^3 \cdot mol^{-1}$. The surface energy U^S of nanodispersed substance is defined by the equation

$$U^S = \sigma_S A, \tag{10.2}$$

where σ_S is the specific surface energy of solid. Consequently, for gold at 298K we have:

$$U^S = \frac{40.035 \cdot 10^{-6}}{r}. \tag{10.3}$$

At $r = 50$, 5, and 0.5 nm the values of U^S make 801, 8010, and 80106 $J \cdot mol^{-1}$. At the stated values of r the total internal energy U^d of gold equals: 6481, 13690, and 85786 $J \cdot mol^{-1}$, and the part of surface energy (U^S/U^d) amounts to: 0.117, 0.585, and 0.934. In other words, the part of surface energy increases from 11.7 to 93.4 % with radius of spherical nanoparticle of gold decreasing from 50 to 0.5 nm.

10.3. From the energy of nanodispersed substance to the energy of aggregates without "bulk" atoms

If the diameter of a spherical nanoparticle decreases, the part of atoms or molecules at its surfaces increases. To estimate a part α^S of atoms (molecules) at the surface of nanodispersed substance, we assume that its molar volume does not depend on the diameter of spherical particles. Then the diameter d of a cross-section of one atom (molecule) of substance can be calculated according to the equation

$$d = \left(\frac{V}{N}\right)^{1/3}, \tag{10.4}$$

where V is the molar volume, N is the Avogadro constant. For gold $d = 0.257$ nm.

If the diameter of a spherical particle of nanodispersed substance is equal to d_n the diameter of its "bulk" part is equal $d_n - 2d$. Consequently, volume v_n of nanoparticle and volumes of atoms (molecules) at its surface v^S are equal to:

$$v = \frac{\pi}{6} d_n^3, \tag{10.5}$$

$$v^S = \frac{\pi}{6} d_n^3 - \frac{\pi}{6}(d_n - 2d)^3. \quad (10.6)$$

Based on the assumption above, the part of atoms (molecules) α^S at the surface of nanoparticle can be estimated by the equation

$$\alpha^S = \frac{v^S}{v}. \quad (10.7)$$

Let us calculate the values of α^S and U^S at 298K for various diameters of spherical nanopartices of gold according to equations (10.7) and (10.3) and further using the equation

$$U^d = U + U^S \quad (10.8)$$

we find the internal energy U^d of nanodispersed gold, taking into account that $U = 5680$ J·mol^{-1}. The results are listed below.

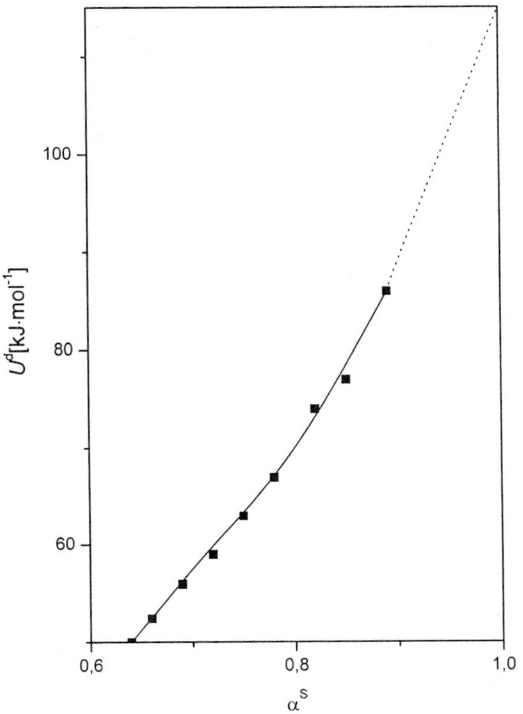

Fig. 10.1. Internal energy U^d of nanodispersed gold versus the part of atoms α^S at its surface (a continuous line). The dashed line corresponds to extrapolation up to α^S = 1.

Diameter of sphere *d* [nm]	α^S	U^d [J·mol^{-1}]
1.0	0.885	85786
1.1	0.849	78504
1.2	0.813	72435
1.3	0.779	67300
1.4	0.746	62898
1.5	0.716	59084
1.6	0.687	55746
1.7	0.660	52801
1.8	0.635	50183

The U^d,α^S dependence for gold at 298K is shown in Fig. 10.1. It is evident that the coefficient α^S increases from 0.635 to 0.885, while U^d from 50183 up to 85786 J·mol^{-1} with reduction of diameter of nanoparticle from 1.8 to 1.0 nm.

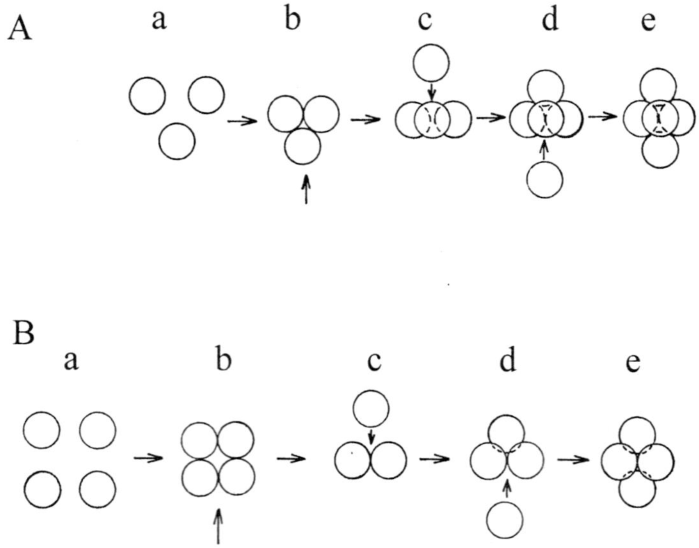

Fig. 10.2. Formation of aggregates without "bulk" atoms from free atoms. A – three atoms (*a, b*) form the triangle which is joined by 2 (*c, d*) more atoms and the dual tetrahedron is formed of five atoms *(e)*. B – four atoms (*a, b*) form the quadrangle which is joined by 2 more atoms (*c, d*) and the six atoms make up an octahedron *(e)*. *(c)*, *(d)*, and *(e)* is a views along an arrow *(b)*.

Extrapolating the U^d,α^S curve up to α^S= 1.0, we obtain the value of U^d= 114000 J·mol^{-1} which refers to aggregates from atoms of gold in which all atoms are at the surface. In these aggregates there are no "bulk" atoms of gold. Thus, when bulk solid gold turns into aggregates without "bulk" atoms its internal energy increases from 5680 to 114000 J·mol^{-1}, i.e. 20 times more.

The spatial location of atoms in aggregates without "bulk" atoms is the one, which one can only suggest at. For example, three atoms (Fig. 10.2, A, *a*) form the triangle *(b)* in which pockets above and below two more atoms (*c, d*) are placed. As a result, a dual tetrahedron develops. In five tops of it there are 5 atoms *(e)*. It is also reasonable to assume that 4 atoms (Fig. 10.2, B, *a*) form a quadrangle *(b)* in which pockets above and below two more atoms are placed (*c, d*). As a result an octahedron in which six tops there are 6 atoms *(e)* develops.

10.4. From the energy of aggregates without "bulk" atoms to the energy of atoms in gas

As stated above, the energy of aggregates composed of gold atoms without "bulk" atoms is equal to 114000 J·mol^{-1}. The energy of gold atoms increases to 374123 J·mol^{-1} [23, 27] in the gas consisting of separate atoms. That is by a factor of 3.28 more. In other words, the difference in the state of atoms of gold in aggregates without "bulk" atoms and those in gas phase leads to the greatest absolute increment in energy (260123 J·mol^{-1}).

10.5. From the energy of atoms in gas to the energy of atoms in solid

Internal energy of atoms of gold in gas $U(g)$ is equal to 374123 J·mol^{-1}. If these atoms form aggregates without "bulk" atoms the internal energy of gold decreases to 114000 J·mol^{-1}, i.e. 260123 J·mol^{-1} less. Such amount of energy is released at formation of units without "bulk" atoms from 5–6 gas atoms.

If these aggregates unite with the formation of nanoparticle with a diameter of 1 nm in which there is 31 atoms, the internal energy of gold decreases to 85786 J·mol^{-1}, i.e. 28214 J·mol^{-1} less, that is by a factor of 9.2 less than at transition of atoms from gas into aggregates without "bulk" atoms. With a further increase in the diameter of nanoparticles from 1 to 100 nm the internal energy of gold decreases from 85786 to 6481 J·mol^{-1}, i.e. 79305 J·mol^{-1} less.

If nanodispersed gold with diameter of particles of 100 nm turns into bulk substance with a diameter of sphere equal 0.0269 m the internal energy of gold decreases from 6481 to 5680 J·mol^{-1}, i.e. 801 J·mol^{-1} less.

Thus, with the state of gold turning from free atoms in gas into atoms in bulk solid the internal energy decreases 368443 J·mol^{-1}. This amount of energy is distributed in the following way:

Process	Energy decreasing [J·mol^{-1}]	%	Number of atoms per particle
Transition from atoms (with measure ~0.25 nm) in gas to aggregates without "bulk" atoms (~0.5 nm)	260126	70.60	1–5 or 6

Transition from aggregates without "bulk" atoms to nanoparticles with a diameter of 1 nm	28214	7.65	5 or 6–31
Transition from nanoparticles with a diameter of 1 nm to those with a diameter of 100 nm	79305	21.52	$31–3.1\cdot10^{7}$
Transition from nanoparticles with a diameter of 100 nm to bulk substance	801	0.22	$3.1\cdot10^{7}–6.02\cdot10^{23}$

As appears, the greatest decrease in energy (70.60 %) occurs at formation of aggregates of 5-6 atoms from gas atoms. The rest 29.40 % fall to rearranging the structure of solid from aggregates without "bulk" atoms to the structure of solid in which surface energy can be neglected.

Conclusion

Nanotechnology encompasses the production and integration of nanodispersed materials into micro- or macrosystems with set properties. Nanosized objects are characterized by properties which differ essentially from those for isolated atoms (molecules) and bulk substances. Neither the atom nor the molecule possess such the well-known properties as electroconductivity, magnetism, color, hardness or a certain melting temperature. Some atoms acquire such properties as a result of a decrease in their energy when aggregates without "surface" atoms are formed. This leads to an increase in their dimensions with formation of nanoparticles in which atoms are partly at the surface, and partly in the volume. The spherical nanoparticle 2.244 nm in diameter formed from atoms of gold melts at a temperature of 400K whereas the melting temperature of bulk gold is equal to 1338K. At the same time the energy of gold decreases from 374123 $J \cdot mol^{-1}$ (atoms in gas) to 41378 $J \cdot mol^{-1}$ (atoms in nanoparticles) and further down to 5680 $kJ \cdot mol^{-1}$ (atoms in the macrophase). The energy of gold decreasing , the rest of its properties (for example, color) also change.

As substance passes from isolated atoms or molecules to bulk substance, a decrease in its energy leads to the change in other properties of substance. Consequently, the question demands careful *experimental* research. A special attention needs to be given to nanodispersed state of substance (the minimum size from 1 up to 100 nm) since this state combines physical, chemical, and biological properties of matter.

When studying energy of nanodispersed substance it is necessary to keep in mind that any change in it is governed by the first and second laws of thermodynamics, i.e. any change in the internal energy of substance is accompanied by a change of its free energy. This mechanism laid into the nature of matter eliminates any theory claiming the possibility to calculate, for example, the free energy of substance. The line of attack is equivalent to the statement that electric resistance of substance can be calculated bypassing the Ohm's law. A reference to the fact that some theory is based on "the first principles" is not a point in favor of replacing the law by a theory. Matter exists according to laws independent of the person and his "first principles". Understanding this inevitability is especially important when creating a foundation for thermodynamics of nanodispersed substance. As discussed above (see chapter 1) the prerequisites for development of nanothermodynamics in another direction have been already created by classical thermodynamics.

Let us just consider the concept of capillary pressure, according to which pressure within a liquid drop can be many times higher than the external one, though being produced without performing the work of compression. Since the internal energy of a liquid increases with the growth of pressure, we come to the conclusion that there is a way to produce energy out of nothing. The thing is that no work of compression is done, which would allow the energy of a substance to increase. But this statement has existed in science since 1806, i.e. for almost two hundred years. There has been no attempt to verify this "scientific" results by measurement. The same can be said about enthalpy (see section 1.1.3).

Nanotechnology will be of great importance in the future. Therefore its scientific foundations should not be based on the available "theoretical achievements". Matter just does not realize them as it exists in conformity with the laws laid in it.

Scientific foundations of nanothermodynamics should be based on nothing but the laws of the existence of matter.

This is a guarantee for achieving success in applying it.

Literature

General literature

1. Clausius, R. Über verschiedene für die Anwendung bequeme Formen der Hauptgleichungen der mechanischen Wärmetheorie, Annalen der Physik und Chemie 125 (1865) 353-400.
2. Dörfler, H-D., Grenzflächen und kolloid- disperse Systeme. Springer Verlag, Berlin, 2002.
3. Adamson, A.W., Gast A.P., Physical Chemistry of Surfaces, John Wiley and Sons, Inc. New York, 1997.
4. Morrison, S.R., The Chemical Physics of Surface, Plenum, London, 1977.
5. Rowlinson, J.S., Widom,B., Molecular Theory of Capillarity, Clarendon Press, Oxford, 1982.
6. Defay, R., Prigogine, I., Bellemans, A., Everett, D.H., Surface Tension and Adsorption, Longmans-Green, London, 1966.
7. Neumann, A.W., Spelt, J.K., eds., Applied Surface Thermodynamics. Surfactant Science Series, Vol. 63, Marcel Dekker, New York, 1996.
8. Buhler, P., Physicochemical Thermodynamics of Substance, Yanus, St.Petersburg, 2000.
9. Buhler, P., Thermodynamics of Substance at High Pressures, Yanus, St.Petersburg, 2002.
10. Buffat, Ph., Borel, J-P., Phys. Rev. A., v.13, 2287 (1976).
11. Popel S.I. Surface phenomena in melts, Moscow, Metallurgiya, 1994 (in Russian).
12. Greeg, S.J., Sing, K.S.W., Adsorption, Surface on Porosity, 2nd ed., Academic Press, Orlando, 1983.
13. Ubbelohde, A.R., The Molten State of Matter. John Wiley and Sons Ltd., New York, 1978.
14. Somorjai, G.A., Introduction to Surface Chemistry and Catalyses. John Wiley and Sons, Unc., New York, 1994.
15. Schwuger, M.J., Lehrbuch der Grenzflächenchemie, Georg Thieme Verlag Stuttgart, New York, 1996.
16. Lagaly, G., Schulz, O., Zimehl,R., Dispersionen und Emulsionen, Steinkopff Darmstadt, 1997.
17. Roco, M.C., Williams,R.S., Alivisatos,P.(Eds), Nanotechnology Research Directions, IWGN Workshop Report,Kluwer Academic Publishers,Dordrecht, Boston, London,2000.
18. Rubhahn, H-G., Nanophysik und Nanotechnologie, B.G.Teubner, Stuttgart, 2002.
19. Fahner, W., Nanotechnologie und Nanoprozesse, Springer-Verlag, Berlin, 2003.

Reference books

20. Wagman, D.D., Evans, W.H., Parker, V.B., Schumm,R.H., Halow, I., Bailey, S.M., Churney, K.L, and Nuttall R.L., The NBS of Chemical Thermodynamic Properties, J. Phys.Chem. Ref. Data 11 Suppl. 2(1982) 1-392.
21. Chase, M.W., Davies, C.A., Downey, J.R., McDonald, R.A. and Syverud, A.N., JANAF Thermochemical Tables, 3rd edn. J. Phys.Chem. Ref. Data 14 Suppl. 1(1985) 1-392.
22. Glushko, V.P.(ed), Termodinamicheskie svoistva individualnykh veshchestv, Vols. 1-4, Nauka, Moskow 1978-85.
23. Knacke, O., Kubaschewski, O., Hesselmann, K.: Thermochemical Properties of Inorganic Substances, 1, 11.Springer - Verlag Berlin, Verlag Stahleisen Düsseldorf, 1991.
24. JANAF, Thermochemical Tables (D.R. Stull et al.) US Dept. Of Commerce/ Natl. Bur.Stand., US Governmt. Printing Office, 1971.
25. Barin, I., Thermochemical Data of Pure Substances, Part. I, II, VCH, Weinheim 1989.
26. Landolt - Börnstein, Numerical Data and Functional Relationships in Science and Technology, New Series, Group 4, Vols.4, 5A, 5B, 5C, 5D, 5E, 5F, 5G, 5H, 5I, 5J, 8A, 8B, 8C, 8D, 8E, 8F, 16, 19, 20A. Springer-Verlag Berlin 1980-99.
27. Elliott, J. F., Gleiser, M., Ramakrishna, V., Thermochemistry for Steelmaking, Reading, Massachusetts, USA, London 1965.
28a. Gmelins Handbuch der Anorganischen Chemie, 8. völlig neu bearbeitete Aufgabe. Verlag Chemie, Weinheim 1950 - 1975.
28b. Gmelin Handbuch der Anorganischen Chemie, 8. völlig neu bearbeitete Aufgabe. Springer-Verlag Berlin 1975 - 1980.
28c. Gmelin Handbook of Inorganic Chemistry, 8 th Edition Springer-Verlag Berlin 1981 - 1991.
28d. Gmelin Handbook of Inorganic and Organometallic Chemistry, 8 th Edition Springer-Verlag Berlin 1990 - 1996. Reference books 28a ÷ 28d have been organized according to elements of periodic system of elements.
29. Handbook of Chemistry and Physics. 83. Edition 2002-2003. CRC Press. Boca Raton, New York, London.
30. L.Haar, J.-S. Gallagher, G.-S.Kell, NBS/NRC Steam Tables, Hemishere Pubshing Corporation, Washington, New York, London, 1984.
31.Landolt-Börnstein. Numerical Data and Functional Relationships in Science and Technology. New Series. Group IV: Physical Chemistry. Volume 16. Surface Tension of Pure Liquids and Binary Liquid Mixtures. Ed.M.D. Lechner, Authors Ch.Wohlfarth and B.Wohlfarth. Springer, Berlin, 1997.
32. IUPAC, Quantities, Units and Symbols in Physical Chemistry, Second Edition, Blackwell Science 1993.

Subject index

The professor Peter Buhler was born in 1935 in the Ukraine (Vyskopol'e, Kherson region) in the family of the Germans – migrants from Western Prussia. His father was repressed and shot in 1937. In 1943 P. Buhler together with his mother went to Germany, whence in 1945 the family was deported to the Soviet Union to live in special settlements for the internee Germans (Kazakhstan, the Urals) till 1956. From 1956 he studied at the Ural polytechnic institute (USTU-UPI, city Sverdlovsk, now Yekaterinburg) graduating in metallurgy in 1961.
Then he worked in the "Giprostal" institute in Kharkov (the Ukraine). When back to Sverdlovsk (Yekaterinburg) he began postgraduate study at UPI and in 1968 was awarded his PhD for a thesis on the theory of metallurgical processes. From this year until 1990 he worked in USTU-UPI sequentially as an assistant, a senior lecturer, and from 1977, after taking his DSc for a doctoral thesis on "Corrosion of metals in oxide melts", as a Professor and the head of department of chemical technology of glass and glassceramics.
In 1990 P. Buhler and his family returned home to Germany and lived in Bonn till 1991.
From 1991 to 2000 he was a Professor of the department of glass and ceramics at the University of Erlangen -Nuremberg. Now he is an adviser at the same department.
Prof. P. Buhler has done a great deal of research on the thermodynamics of interaction of oxide melts with metals and gases. A series of articles on thermodynamics of interaction between oxide melts and gases has been published in the Russian journal "Physics and chemistry of glass". In 1998 a P. Buhler's book entitled " Wechselwirkung zwischen Glasschmelzen und Gasen " (Riga, Verlag "Saknes") was published. Then the following books by P. Buhler were published in the "Yanus" Published House (Russia, St.-Petersburg): "Physicochemical Thermodynamics of Substance" (2000), "Fizikokhimicheskaya termodinamika veshchestva" (in Russian, 2001), "Thermodynamics of substances at high pressures" (2002), "Termodinamika veshchestva pri vysokov davleniy"(in Russian, 2002).